Roderick H. l

# <u>The Physics of Encounter</u>

## *Toward a Theory of Consciousness*

*Does consciousness exist throughout the universe?*

Based on scientific investigations of anomalous phenomena, this book addresses a number of related questions.

Does it explain the reports of encounters with "aliens", apparitions, disembodied minds?

Does it explain the ghost-like forces that cause Recurrent Spontaneous Psychokinesis (RSPK) = "poltergeist" events?

Does it explain why experiments have shown that prayers can heal?

Does is provide evidence that the mind survives the death of the body?

Yes, it does.

It points the way toward an increasingly fruitful dialogue between science and religion.

It points the way toward a more firmly grounded assumption that encounters with the unknown during our lifetime arc but a foretaste of our ultimate encounter with the Universal Mind.

Order this book online at www.trafford.com
or email orders@trafford.com

Most Trafford titles are also available at major online book retailers.

Note for Librarians: A cataloguing record for this book is available from Library
and Archives Canada at www.collectionscanada.ca/amicus/index-e.html

Printed in Victoria, BC, Canada.

ISBN: 978-1-4269-1089-0

*Our mission is to efficiently provide the world's finest, most comprehensive book publishing
service, enabling every author to experience success. To find out how to publish your book, your
way, and have it available worldwide, visit us online at www.trafford.com*

*Trafford rev. 9/23/2009*

**PUBLISHING** www.trafford.com

**North America & international**
toll-free: 1 888 232 4444 (USA & Canada)
phone: 250 383 6864 ♦ fax: 812 355 4082

# Table of Contents

## Part One

**What is the physical reality of consciousness?**

# Part Three

**Crucial issues in theoretical physics**

# Acknowledgements

Back in 1950, when I enrolled in college, such concepts as quantum vacuum fluctuations and oscillating strings were not yet in the textbooks on physics. The quirky particles that physicists decided to call quarks had also not yet been introduced as conceptual constructs. I recognized their pivotal role in what has become the subject matter of this book while I was working as a journalist in Germany. After my retirement, I was able to visit the scientific conferences that helped me bring my project to fruition. I am grateful to Prof. Hal Puthoff for the detailed description of his interpretation of quantum vacuum fluctuations and their relevance to my own approach. Encouragement and feedback came from many members of the Society for Scientific Exploration, in particular from Prof. Robert Jahn, Prof. Henry Bauer, Prof. Garret Moddel, Prof. Hans-W. Wendt, and Dr. James Beichler.

The nitty-gritty details of how to apply the laws of physics to the study of anomalies were the subject of many lively discussions I have had in my hometown of Berlin with the physicist and parapsychologist Dr. Wilfried Kugel. As shown in chapter six of this book, the "logical" interpretation of empirical evidence in this field is often a matter of hot scientific dispute. I am indebted to Dr. Kugel for helping me pursue my own thoughts, independent of his conclusions, on the basis of published research, about which he provided helpful information.

I am grateful to Dr. Edward Lantz for his supportive reaction to my metaphorical images of quantum vacuum bubbles containing the qualia of consciousness. They correspond to the principle used in his projects involving total sensory immersion in wrap-around screens. On the subject of UFOs, I was encouraged by the support of the geophysicist Dr. Oliver Stummer. Prof. Leo Ferrera of MUFON-CES provided me with many useful articles and sources of information while I was attending his workshops. Reinhard Nühlen of DEGUFO was very helpful in alerting me to various conferences and possible contacts regarding UFOs and other anomalies.

Many details in this book are owed to my encounters with a broad spectrum of scientific interests. I am grateful for the conversations with Dr. Rosemarie Pilkington about her research on séances, with Prof. William Roll about Poltergeist phenomena, and with Dr. Roger Nelson about his Global Consciousness Program. Thoughts shared with Dr. Lawrence Fagg, Dr. Robert Brueck, and many others, have confirmed my views about the spiritual aspects of scientific research.

# Preface

Will scientists be able to build computers that have "a mind"? Does it take a brain to create consciousness? The Physics of Encounter suggests that a mind can be regarded as a physical system that processes information and *feels* something while doing so. How does physical reality produce feelings? What are mind events? Physicists are unable to explain the paradoxical relationship between the measurable effects of matter and the inaccessible "inner" reality of a mind that is aware of its own existence.

The Physics of Encounter proposes a geometric model for describing the unobservable processes that create the events experienced in the mind and also create the elementary physical events in space and time. The proposed images of the interaction between mind and matter touch upon deeply personal philosophies and religious views. This book, therefore, was written for broadly interested readers of all backgrounds. Where I have quoted from scientific publications, I have explained the concepts that may not be familiar to everyone. Theoretical physicists should find the book worthwhile because it suggests answers to some vexing questions within their field.

The puzzles addressed in this book involves more than the "normal" events that occur when physical reality affects the mind. Why is it that just *thinking* something can influence events outside the mind? Two radically different kinds of reality seem to be "entangled". What evidence is there that the mind can exist outside the brain? Is it possible that our consciousness survives the death of our physical body?

Part One of this book describes the basic features of the Physics of Encounter. Part Two shows that this model can shed some light on the details of "anomalous" events that the currently known laws of nature cannot explain. Part Three shows that the proposed scenario is in accord with a broad range of conceptual constructs currently used by physicists.

For easy cross-reference, I have identified quoted arguments when they are repeated in a different context. This applies, for example, to the argument by Jahn and Dunne (3/8) about the "generous use of conceptual metaphors". The numbers in parenthesis indicate that this argument was first quoted and elaborated in chapter three, reference note number eight.

The proposed model of mind/matter interaction shows that appropriate metaphors are a first step in developing a new theory. The Physics of Encounter provides metaphorical descriptions of assumed, unobservable processes. Needless to say, it does not claim to be a full-fledged theory. In one important respect, however, the metaphorical model presented in this book does what a good theory should do: it proposes experiments for verifying the assumptions. The proposed experiments are described in chapter twenty-two.

# PART ONE

## What is the physical reality of consciousness?

## *Chapter 1*

### Open questions about the laws of nature.

Can physics help us understand telepathy? How can the mind of one person exert a direct effect on another person's mind? What about the effect of mind on matter? Can the human mind bend forks, or were the demonstrations of this purported feat just a hoax? Whatever the answer, there's strong experimental evidence that the mind does influence elementary physical processes. Nobody knows how this is possible. Experiments that provide facts are one thing. Educated guesses how to interpret them is something else. Do some people possess a mysterious power to heal because their mind enables them to focus their bioenergy on the ailments of others? If so, what is that energy? What about UFOs and the "aliens" who reportedly pilot them to earth from far-away regions of our universe? Should we take all of this as evidence that there are conscious minds with extraordinary abilities that are "not from this world"? The answers to these and similar questions are open to debate.

Not all of this debate is scientific. Strong emotions and attitudes come into play. Heated arguments are exchanged between two highly polarized groups. There are the stern "debunkers" who regard much of what is published on this subject matter as myth and sensationalism. On the other side of the barricade are those who ferociously defend what are, at best, controversial beliefs.

The main purpose of this book is to show that much of what science cannot (yet) "prove" is worth a closer look, based on an enlarged understanding of the laws of nature. I would like to stimulate a feeling of intellectual adventure and of reverence for the majestic possibilities of the mind as it interacts with physical reality and triggers the amazing events that are called "anomalous". The Physics of Encounter suggests a conceptual model of the processes involved. It is supported by a broad array of experimentally verified facts.

Secondly, the book points to some of the good reasons for a healthy dose of skepticism regarding claims about events that seem to violate the laws of nature. The arguments presented here, therefore, also highlight the need to "debunk" what is pure hype and irrationalism.

Thirdly, the description of the views on the above issues provides some eye-opening insights into the battles fought in the scientific establishment. Keen minds trained in logical reasoning are embroiled in quarrels over the interpretation of facts, over scientific methodology, and the spiritual aspect of efforts to understand the universe in which we live.

This book describes the role of a *hidden source of energy* in producing anomalous events. When UFOs are observed, unusual magnetic disturbances make compass needles gyrate wildly, stalls automobile engines, and garbles radio reception. In published investigations we read that at locations where UFOs appear to have landed, the earth was scorched and plants had aged prematurely. Many of the sightings were reported by airline pilots. Radar operators at airports confirmed that they saw blips on their screens that matched the location of the sightings. In several countries, the military was sufficiently alarmed to dispatch fighter planes.

The scientific debate over the origin of this hidden source of energy, and what it can do, spills over into the domain of *religion*. The energy creates our consciousness, which we gratefully experience as something radically different from what is regarded as the blind and unfeeling reality of material objects. Our consciousness fills us with awe because it lets us sense the immense possibilities of a mind that resides outside the limitations of space and time. On the worldly plane, telepathy shows that a mind can overcome these limitations. Who is to say whether the human mind is the only mind that is active in this universe of ours? Do some of the mysterious events involve *disembodied minds*?

Evidence is accumulating that consciousness can occur without any functional involvement of the brain. Much of the evidence comes from patients who were brought back to life after they had what seemed to be a fatal accident or heart attack. They were brain-dead and their heart had stopped beating. Nonetheless, the mind of these patients was keenly in action while doctors were laboring to revive them. According to those who were at the operating table and spoke to their patients afterwards, and other investigators of these *Near-Death-Experiences (NDEs)*, the patients reported that their mind floated above the operating table and enabled them to watch what was happening.

From this unique vantage point, they could see things that would have been totally outside their field of vision while lying on the operating table. Their brain was not functioning, but they could hear what was being said. When they quoted what they had heard, those who had done the talking were stunned that it corresponded precisely to their recollection. As reported by revived patients, their mind was pulled into the depths of a beyond where it received messages from a bright and loving source of light. Others reported encounters with images of loved ones who had died. These images communicated wordlessly, as if by *telepathy*.

There are strong similarities between such experiences and those described, in a totally different context, as encounters with the occupants of UFOs. The latter are experienced as encounters with alien beings. The "*aliens*", too, communicate by telepathy. In many consistent reports from those who say they encountered them, the aliens expressed deep concern about the future of mankind, in particular about environmental devastation. In some respects, they act like the angels described in religious literature, admonishing us with wisdom and compassion. They seem intent on keeping mankind from coming to harm. In other reports, however, these aliens are described as coarsely indifferent to the feelings of individual humans as they pursue their goal of investigating the nature of life on earth. (Sinister plots to conquer our planet are not usually reported. They are the plots in science fiction movies.) Psychiatrists have examined many of the people who described such encounters - engineers, architects, housewives - and found that they were perfectly normal individuals.

The encounters with aliens or with loved ones during an NDE all involve images of *disembodied minds*. NDEs usually instill a strong belief that our mind will survive the death of our body. The Physics of Encounter supports that belief. It provides argument for the *pantheistic* view that God and the universe are one and the same, and that there may well be a cosmic consciousness.

That is reason enough to look for an alternative interpretation of the event that astrophysicists call the *big bang*, the presumed explosion that brought our universe into existence. Was that an act of creation? Or is it more logical to assume that the universe did not require an act of creation because it has always existed and will always exist? There are plausible cosmic scenarios that describe this possibility.

The Physics of Encounter suggests that, in our universe, all of reality is so interconnected that events occurring elsewhere, and/or at other times, can directly influence our minds here and now, and vice versa. There is a growing body of evidence that the effects of the mind on physical reality are exerted *instantly*. Physicists have begun to grapple with what seems like a logical impossibility: that events or changes in an unobservable reality can occur without the passage of time. As argued in this book, this occurs at an elementary level of reality that brings forth the physical reality outside ourselves (*matter*) as well as the reality we experience inside ourselves (*mind*). Both mind and matter are what physicists have called *emergent realities* arising from processes that occur below the level of observable physical events.

The processes occur in what physicists have called an *unobservable substratum* of ordinary reality. Hal Puthoff has described it as the *pre-manifest reality* of quantum vacuum fluctuations.(1) These fluctuations of non-measurable energy pervade all of space, even the vacuum in so-called "empty space". The deeper and more fundamental source of energy is not like a separate layer of reality. It is intertwined with the ordinary reality that is "out in the open". The paradox of timeless events in the "other reality" has become the subject of many scientific articles. The physicist John Briggs put it this way to the science journalist Thomas de Padova: "Time does not enter the picture until a quantum system interacts with its surroundings."(2)

A quantum system contains interacting elementary particles. Each particle consists of a certain amount, or quantum, of energy. Quantum systems are the building blocks of all observable objects, including the human brain. In the quantum vacuum fluctuations of space, there are no quantum systems.

The Physics of Encounter describes how certain mind events are brought about by processes outside the human brain. They can occur in *disembodied minds* because the effects of events in the timeless reality underlying both mind and matter become instantly interconnected in large volumes of space. In the hidden reality of quantum vacuum fluctuations, the cause of an effect can jump across time as we experience it. Theoretical physicists have argued that the energy that creates elementary particles can influence not only future events, but also events that, for us, lie in the past.

In the world we observe, and in our minds, however, time only flows forward. Why is this so? This book will describe the processes involved.

My use of the expression "flow of time" will raise some eyebrows among mainstream physicists. For good reasons, they regard time as a dimension, not as a "flow" of something. The proposed metaphorical model does not contradict conventional scientific wisdom, but offers additional food for thought. To understand the nature of time, imagine a clock at the edge of a rapidly rotating disk and a stationary clock at the center of the disk. When you remove both clocks, you will notice that the one that was whisked through space has ticked more slowly. That's the *relativity of time*. It makes you wonder. Something happened inside the clock that did not happen inside your mind. The Physics of Encounter suggests that time is not just a dimension, but also a process.

If you could travel through space at the speed of light, time, for you, would slow to a stop. The elusive reality of light and the counter-intuitive conclusions of relativity theories are a challenge to all critical minds.

> *There was a young lady named Bright,*
> *whose speed was far faster than light.*
> *She set out one day,*
> *In a relative way,*
> *And returned on the previous night.*
> A. H. Reginald Buller

This is not possible, of course, not even theoretically. No physical body can travel at the speed of light, and the so-called *"arrow of time"* points both ways only in the world of elementary particles. The Physics of Encounter proposes a model for the processes involved.

Both time and space, in the proposed geometric model of mind/matter interaction, arise from *quantum vacuum fluctuations*. As interpreted here, a quantum vacuum contains something that is not yet anything measurable. It contains the invisible stuff of which mind and matter are made. It makes time flow. Expressed in terms of the images described in this book, experiments have shown that time can flow at different speeds. Putting a clock on a rotating disk is only an imagined experiment. Using a term introduced by Albert Einstein, based on the German word for thought, physicists have called the experiment a *Gedankenexperiment*. It illustrates that we "go with the flow". We are inside the flow of time we experience and outside the time of the clock that ticks more slowly. What seems to occur in separate worlds, however, inside our mind and outside our mind, is no longer separate at a higher level of understanding. The flow of time is like a river in which water swirls around rocks at different speeds as the current heads downstream.

In the above "Gedankenexperiment", the clock in which time is slowed is like the mind of another person. We cannot look into it to see what is happening. Something quite similar happens in the mind of someone observing a UFO, or experiences what seems to be an abduction by its alien occupants. The expression *"missing time"* is often used by people who had the deeply unsettling experience of encountering this strange, inexplicable reality during an *altered state of consciousness*. The experience, in their mind, was very intense and seemed to last only a few minutes. When they found themselves back in the real world, they noticed that they had been "away" for a much longer period of time. In some instances, several hours had passed.

The intensity of such abnormal experiences is very similar to *Near-Death-Experiences*. Both are often described as "far more real than ordinary reality". The Physics of Encounter suggests that the intensity of an experience is related to the unobservable processes that influence the density of matter. This ties in with the *theories of relativity* which state that when time is slowed at high velocities, space shrinks and the density of matter increases. In the proposed model of mind/matter interaction, the mind that is flooded with abnormally intense experiences does not move through space like the clock at the edge of a rotating disk. In a manner of speaking, we could say that space moves through the mind. Space fluctuates. The *quantum vacuum fluctuations* that give rise to the reality of space also influence our mind.

I have already mentioned that my arguments will raise some eyebrows because they seem to conflict with current theories, which describe the relativity of space and time in terms of tilted *coordinate systems*. The angle of the tilt is determined by the velocity of one system relative to the other.
A system of coordinates is like a picture on a wall in front of you. The picture has two coordinates, which correspond to the vertical and horizontal edges of the picture. If there is no wall behind the picture and its top edge is tilted away from you, the picture you see is distorted. It is compressed. The space occupied by the picture you see becomes smaller. Space "shrinks". That corresponds to the increase in the density of matter mentioned above. Tilting the picture by 90 degrees relative to your line of vision, as described above, will compress the picture into a horizontal line. If the picture is tilted

so that one of its *vertical* edges moves away from you, this distorts and compresses the picture in a different way.

The picture is a coordinate system that has only two coordinates. The equations of relativity theories describe the relative shortening or lengthening of *four coordinates*: the dimensions of space and of time. One of the coordinate systems that tilts, relative to the other, is the one that describes, in mathematical terms, the relative motion of the observer. The model of mind/matter interaction proposed here describes how the "tilt" ties in with the quantum vacuum fluctuations of space. How should we visualize these fluctuations? They create the reality that exists outside ourselves (matter), and the reality we experience inside ourselves (mind). To describe them is not an easy task. The next chapter will show what difficulties are involved.

*NOTES*
(1) Puthoff (August 2002), *Personal communication*
(2) de Padova (1998), p. 29

## *Chapter 2*

**Physics and metaphors. "Visualizing" invisible forces. Waves and bubbles. The pull of gravity and the push of waves. Points of encounter and zero-point energy.**

To visualize the unobservable processes through which our consciousness "tilts reality", so to speak, we need to be clear about the distinction between the reality inside the mind and the "outside" reality of matter events. Let's say you are deep asleep in your bedroom. During that time, you are not aware of your own existence. Your consciousness has taken "time off". Let's say the sun has just gone up and a dog starts barking outside. This wakes you up. You see the dog through the window. Now there is a reality inside your mind that was not there before. You see and hear the dog and probably feel a little irritated to be awakened so early. The dog was out there all the time, but now there are *two realities*: the dog out there and the mix of events in your mind. The mix includes the image of the dog, the sound of the dog's barking, and the irritation you feel. The reality of your consciousness has popped into existence.

What happened? Your brain has revved up and re-created the normal waking state of your mind. It has created images, sounds, and feelings. All of that is an *additional* reality, which is not identical with the dog outside your window and yet somehow intertwined with it.

The physical reality of the dog is a structure of interrelated atoms. So is your brain. When it creates your consciousness, it generates electrical events, measurable energy fields, wave patterns that can be observed on an oscillograph. None of those observable (measurable) events are identical with your consciousness. What happens inside your mind is not accessible to anyone outside your mind. Your feelings, needless to say, are strictly your own. The same is true for the images in your mind and the sounds of which you are aware. No matter how deeply scientists probe into the interconnected atoms of your brain, they will not find the replica of a dog there. How big is that dog in your mind, anyway? It is not a measurable reality.

The events we need to understand, in terms of unobservable physical processes, are events in the *mind*, not in the *brain*. The observable physical reality of the dog exists *outside* your mind. It is the reality that the German philosopher Immanuel Kant called "*the thing-in-itself*". Kant argued that this reality is unknowable. All you can possibly know is the world of your inner experience. Sure, you see the dog. But that's because images arise in your mind. What is the dog really like? It consists of atoms. And atoms, the building blocks of "the thing-in-itself", are mostly empty space.

In a physics course I took as a student in the 1950's, I learned that we could picture the structure of an atom as a miniature solar system, because the atomic nucleus is like a sun surrounded by planets, i.e. by electrons, which travel around the nucleus in orbits. To visualize the dimensions of subatomic particles, we were told that if an atom were the size of the Notre Dame Cathedral in Paris, the *nucleus* would be no larger than a pea and the *orbital electrons* would be like specks of dust. How does so much empty space become the building block of something we can see and touch?

What holds the atom together? How should we imagine its surface? Instead of the walls of a huge cathedral, there are just specks of dust. Besides, some invisible force inside the atom must cause electrons to move in orbits, and to change orbits in what are called *quantum jumps*. We can picture an atom, but we can't picture that force. How do you visualize something that is invisible? Take the influence of gravity, for example. We cannot see gravity. But it is "out there", everywhere in the universe. Einstein said that gravity is not a "force" in the usual sense. Gravity, he said, is just the "geometry of space" that influences the motion of objects.

This book is about the geometry of invisible, elementary processes. They create the elusive substance of elementary particles and the events we can see (moving objects) and they also create the events we experience in our mind: consciousness. John Wheeler hit the nail on the head when he said: "There is nothing in the world except empty, curved space. (...) Physics is geometry."(1) Wheeler died in 2008 at the age of 96 and was hailed as "one of the most influential theoretical physicists of the past century." At the university of Princeton he gained a reputation as an "expert on the unimaginable".(2)

The Physics of Encounter proposes metaphorical images to describe what we cannot see. It describes consciousness in terms of an invisible physical reality. The images suggest that gravity and mind have much in common. Gravity is a mysterious influence. It is like a force of attraction. Throughout our universe, the pull of gravity makes it seem as if objects are tugging at each other across empty space. Big (massive) objects exert a stronger pull than smaller (less massive) ones. The pull of the big object is like that of an invisible fisherman reeling in his catch on an invisible line. The gravitational attraction of our earth makes objects fall to the ground.

In some respects, gravity is like light. Both are a reality "out there" that we cannot touch. But there is a basic difference. Gravity affects the motion of our body. Light affects the experiences in our mind. We can rely on the pull of gravity if we decide to cool off by plopping into a swimming pool. Light, on the other hand, does not pull us. It comes toward us.

Atoms send out particles of light called photons. The energy of a photon is a puzzling kind of either/or reality. A photon can hit a surface like a particle. Then we know where it is, or more precisely: where it *was* at the instant it hit the surface and created an effect. As soon as a photon makes an impact, it is no longer at that point. It is out of sight, so to speak, because it is replaced by a wave, like a pebble that was dropped into a pond. The pebble sinks, but now there are waves.

The waves expand outward from the point where the pebble disappeared. They gradually flatten out, but if we imagine them as ocean waves splashing against the walls protecting a seaport, we can see that they are not just an up-and-down motion within a surface. They push against the walls of the seaport. In the world of elementary physical reality, there are two kinds of forces. Some push, like waves, and some pull, like gravity.

The waves, the pebble and the pond are similes, figures of speech in which one thing is compared to another. In this case, they give us some idea of what photons are, or rather: what they do, even though we cannot see them. For the sake of simplicity, I will not use the word simile in this book, but the better known word "metaphor", although the dictionary tells us that the meanings are not quite the same. *Metaphors* are words or expressions that clarify matters, but should not be taken literally. When physicists speak of electron orbits, for example, they are using a metaphor.

The word *electron orbit* is a metaphor because electrons do not travel in orbits the way planets do in a solar system. We cannot see electrons moving around the nucleus of an atom. In fact, we cannot see them at all when they are inside an atom. We can only make educated guesses about what is happening inside an atom. The orbit of an electron is an array of points around the atomic nucleus where electrons can be assumed to exist for fleeting moments, sometimes here, sometimes there. The same goes for the word *quantum jump*. We cannot see an electron making the jump. First the electron is in one orbit, then it is suddenly no longer there. It simply disappears and instantly re-appears in a different orbit. It skips over the space in between.

A change in orbit changes an electron's characteristics. What are they? As an elementary particle, an electron does not have very many. Its energy is one of them. It changes when an electron makes a quantum jump. If all of a particle's characteristics radically change at once, which can happen, wouldn't the particle lose its identity? Is an electron making a quantum jump still "the same" electron? Instead of saying that a particle jumps from here to there, shouldn't we say that it ceases to exist and that, at the same instant, a different particle surfaces elsewhere? The world of quantum jumps is a conceptual minefield. It can blow away certainties about the nature of reality. That is not necessarily a bad thing. Some former certainties need to be discarded to arrive at a better understanding of invisible forces.

Another elementary characteristic, also a metaphor, is called *particle spin*. Here, too, it's the same story. We cannot see what we want to describe. We cannot verify that something actually rotates. The logical justification for saying that some particles have spin is the fact that the observer of a particle may suddenly be confronted with a surface of the particle that has a different characteristic than the surface previously observed. Then, continuing to look at the particle, the observer sees the same surface again. The conclusion seems obvious: seeing different surfaces of the same thing while keeping an eye on it must mean that it rotates.

But it is not that simple. Metaphorically speaking, it is not possible to keep an eye on the world of particle spin and quantum jumps without blinking. In that kind of a world, what is a surface, anyway? You guessed it: it is a metaphor. A *particle* is not a concrete thing with a surface we can see or touch. It is more like a bundle of intertwined processes, something dynamic, not static.

In that kind of a world, a *surface* is also something dynamic, in constant change. It is not like the flat surface of a pond into which we can drop a pebble. In one way, though, the dynamic surface in the world of invisible forces and the placid surface of a pond are similar. They both separate two different kinds of reality. The surface of the pond separates it from the air above it. The surface is "there", all right, but it exists only if two other things exist: the thing that is the pond and the thing that is *not* the pond (the air above it). All things have a surface. If we consider the whole universe as one thing, could we say that the universe has a surface? It would be a surface that separates something from nothing. This is not like the question about how many angels fit on the point of a pin. In theoretical physics, the question about the nature of "nothing" inside a world that is filled with something is not a meaningless question.

In the world of particle spin and quantum jumps, a surface separates two different invisible forces. That makes it pretty tough to visualize the surface, i.e. to come up with suitable metaphors. Two metaphors that come to mind are the pull of gravity and the push of waves. In a world of invisible surfaces, these metaphors make sense only if we put them into the context of another invisible "something": a *point*. Like a surface, a point is nothing by itself. It exists only as part of a geometric image, or configuration, that includes something else.

In theoretical physics, there are point-particles. A *photon* is such a particle. It makes an impact at the point where it hits matter. Since matter consists of atoms, does this mean that a photon hits the surface of an atom? Photons are emitted from atoms and are absorbed back into other atoms. They "make waves" while doing so. Let's go with the image that an atom is mostly empty space in which invisible forces influence the electrons that exist at uncertain locations around the nucleus. If we visualize all of these possible locations in their entirety, they are an array of points surrounding the nucleus. Since electrons can be anywhere within an orbit, the points are located next to one another. Geometrically, that corresponds to a closed, more or less spherical surface. It is not like a membrane with a certain "thickness". The surface is there because two opposite forces are there.

A photon that is emitted from an atom hits the surface of the atom that absorbs it, like a pebble dropped into the surface of a pond. The impact makes waves in the electron orbits and in the space across which electrons make quantum jumps. The waves are changes in the array of points marking the possible locations of electrons in the atom. Since there is no wall that separates the inside of an atom from the outside, the waves inside atoms spread out into the space outside atoms, and vice versa.

If we picture space as a sea of invisible waves, an atom is like a natural harbor. Outside the harbor, there are strong winds and ships traveling at full speed. Think of the choppy waters when ocean waves are pushed against one another by gusting winds blowing from different directions. For good measure, picture the waves made by ships criss-crossing the surface of the ocean. All of these waves interfere with one another and create complex patterns of ups and downs. The winds, in this image, are invisible forces. The ships symbolize physical objects. At the center of the metaphorical harbor, things are not as calm as they appear to be on the outside of the atom. The *atomic nucleus* is like a caldron of boiling water, bubbling over, vibrating violently and passing on those vibrations to the space around it.

To continue the description of events in the world of atoms and particles, it is necessary to switch metaphors. That's because the surfaces in that world are not like the surfaces of ponds and oceans. These are more or less flat. The surfaces of atoms and particles, on the other hand, enclose something. Very broadly speaking, these surfaces are like the surfaces of bubbles. Atoms and most particles, as interpreted here, have many surfaces that are like interconnected bubbles. Not static bubbles, but like the soap bubbles that expand while children are blowing them up. The waves in space, outside atoms, are also like the surfaces of expanding bubbles, and these bubbles also enclose something. Metaphorically speaking, that "something" is like the breath that expands the soap bubbles mentioned above. It is an invisible force rooted in the nature of our universe. In later chapters, the nature of that "breath" will, of course, be explained in more detail.

To make these metaphors fit the elementary reality described in physical theories, we must take into consideration that there are waves on the surfaces of the bubbles, and these waves are also like the surfaces of expanding bubbles. Since there are waves in the surfaces of these bubbles, too, the surface of a bubble consists of layers of bubbles which get smaller and smaller toward the outer layer of these metaphorical bubbles.

The Physics of Encounter suggests that the smallest possible bubble is not of the same "stuff" as the bigger bubbles. That's because the smallest possible bubble cannot have a surface consisting of bubbles. If it did, it would not be the smallest possible bubble. Instead, the smallest possible bubble must have a surface consisting only of *points that will expand into bubbles*.

In the proposed scenario of mind/matter interaction, this point corresponds to what physicists have called *zero-point energy (ZPE)*, or vacuum energy. In this scenario, the quantum vacuum fluctuations that exist throughout space are expanding and collapsing bubbles. Each of these bubbles is a quantum vacuum. I will therefore call it a *QV-bubble*.

When a QV-bubble expands, its surface interacts with the invisible stuff of which other bubbles are made. This involves a *stage of uncertainty*. The surface interacts with another surface only if both surfaces *resonate* with each other. If this is not the case, the surfaces pass through one another and continue to expand. The resonance is a process that occurs at what I have called a *point of encounter*. Surfaces can resonate with one another because every surface is like a complex pattern of waves that are superposed on one another. If this resonance occurs at a point of encounter, the surfaces involved in the process collapse. The process creates new QV-bubbles, which expand from points of encounter. The details of this process are described in the following chapters.

Paradoxically, the process creates certainty from an initial uncertainty. The *unobservable* elementary reality creates the *observable* higher-level reality characterized by the certainty of cause and effect encapsulated in the laws of nature. As Isaac Newton showed, for example, we can count on the laws of nature when an apple falls out of a tree. In this case, it is the law of universal gravitation.

Originally, the motion of objects was the domain of what is called Newtonian physics. Einstein enlarged Newton's theoretical framework by looking at the motion of objects relative to the motion of the person observing the motion. The observed motion of a specific object at a specific moment, he found, is not something two different observers would agree upon if their own motion is not the same, relative to the moving object. Motion is measurable in terms of the time required to travel a certain distance through space. But Einstein showed that time and space are not immutable realities that exist on their own, independent of the observer.

Einstein's theory of relativity is a monumental landmark in the efforts to understand the laws of nature. But it is not the "whole picture". Just like Newtonian physics, it illuminates only certain aspects of reality. More problematic is the fact that the relativity of space and time, as described by current theories, does not fit into the framework of an equally fruitful theory called quantum mechanics. It describes the interaction between elementary particles. Einstein's equations pinpoint the precise location of a gravity effect. The theory of quantum mechanics uses a different set of images, or metaphorical representations. The "particles" are metaphors. A particle is a measurable amount or quantum of energy, but it is not a graspable something. It is more like an intertwined flow of energies at a location that cannot be precisely pinpointed in space and time. The location is shrouded in the above-mentioned *uncertainty*.

To reconcile the contradictions between quantum mechanics and relativity theories, a number of physicists have come up with the theory of *oscillating strings*, or oscillating thrusts. Opinions are divided whether string theory does what it is supposed to do. The Physics of Encounter proposes an enlarged interpretation of these metaphorical strings. This interpretation describes how the thrusts inject vacuum energy into the above-mentioned QV-bubbles. The bubbles expand and collapse as their surfaces interact at points of encounter. In the proposed geometric scenario, the quantum vacuum fluctuations that result from these encounters are the unobservable reality from which all observable reality arises. That includes the events described by quantum mechanics as well as Einstein's "geometry of space". It also includes the process that underlies the emergence of consciousness from physical reality. It is a process that allows the mind to influence what is observed. The pivotal role of oscillating strings in this process is described in the next chapter.

*NOTES*
(1) Misner and Wheeler (1957), p. 526
(2) *Der Spiegel* (2008), p. 200

# *Chapter 3*

**Oscillating strings and the ingredients of consciousness. Intertwined realities. "There is more to time than just a dimension." The initial void: creating something from nothing.**

Two metaphorical images dominate the geometric scenario proposed by the Physics of Encounter to describe the unobservable physical origin of consciousness. One is the interaction between QV-bubbles that results in *quantum vacuum fluctuations.* The other is the interaction between *oscillating strings.* Strings come into play as the smallest possible elementary reality in the expanding surface of a QV-bubble. That smallest possible entity must be smaller than a bubble, but it must, logically, be larger than a point. A point by itself is nothing. A point has no dimensionality. But two points, interconnected by an invisible force that oscillates between them, would fit the bill.

The expanding spherical surfaces of QV-bubbles consist of countless oscillating thrusts that push outward, in the direction of the expansion. The thrusts also interconnect the points within the spherical surface, oscillating at right angles to the direction of the expansion. Based on the interrelated images of QV-bubbles and oscillating strings, the Physics of Encounter clarifies three important aspects of mind/matter interaction that have not been satisfactorily explained by current physical theories.

1) The scenario of spherical surfaces expanding from QV-bubbles provides a geometric model of the distinction between *inside* reality and *outside* reality. The events that occur inside a person's mind are not observable by anyone else. They are *subjective* reality. The events that occur outside a person's mind are *objective* reality. Mind events are enclosed events and therefore unobservable by others. They occur on the inside of the spherical surface. Observed events occur on the other side of the spherical surface.

2) In this scenario, the elapse of *time* can be defined as a process that involves both inside and outside reality. The process occurs as interacting spherical surfaces expand and collapse. The subjective reality of time is the continuously changing mix of the elementary ingredients of consciousness. The objective reality of time is the decay of elementary particles. The role of oscillating strings in the creation and interaction of particles and in creating

the non-particle reality of consciousness is one of the key issues examined in this book.

3) In the proposed scenario, the *location of the observer*, relative to the observed reality, can be pinpointed in space and in time. That is important because seeing something involves photons, the particles of light. They impact on atoms and are emitted from atoms. An act of observation needs to be understood in terms of events in the world of elementary particles and the forces acting between them. Those events include the observer and the observer's influence on what is observed. In the Physics of Encounter, the location of the observer is defined in terms of expanding spherical surfaces and the point of encounter between these surfaces. QV-bubbles containing the elementary ingredients of consciousness are created at points of encounter. The observer's location is where the observer's consciousness is continuously re-created from one point in time to the next.

The motion of the observer (a continuous change in location) is a pivotal concept in the theory of relativity. If the observer moves at high speed (in an accelerating spaceship, for example), this changes the reality in which the observer is located. Time is dilated and space contracts. More precisely: the motion changes the interaction between the moving physical system and the space through which that physical system moves. The physical system is the spaceship and everything in it, including the observer. The change is "relative". The interaction differs from the reality observed by someone who is not in such high-speed motion. The contraction of space increases the density of all physical objects in that space. The objects acquire more mass and become heavier. The dilation of time makes clocks tick more slowly and also retards all biological processes. The body of a high-speed space traveler ages more slowly.

The Physics of Encounter shows that the dilation of time increases the *intensity* of each experienced moment. During an altered state of consciousness produced by an anomalous event such as the observation of a UFO or a Near-Death-Experience (NDE), time is slowed through a process that corresponds to accelerated motion through space. The location of the observer, in this scenario, should not be equated with the location of the observer's physical body. It is the location of an unobservable process that influences the atoms of the brain. Usually, that process takes place in the brain. But during an NDE, for example, many people, while clinically dead during a brief period of time, have a so-called "out-of-body" experience.

When they were brought back to life after a heart attack or a serious accident, their descriptions of this experience were remarkably consistent.

The term "observer", in theoretical physics, can be equated with the *consciousness*, or mind, of the observer. An act of observation, as defined here, occurs in the nucleus of an atom. It is a process that resonates, through quantum vacuum fluctutions, with countless processes in the nuclei of other atoms. A mind event, therefore, can instantly influence events at other locations in space. In psychokinesis, for example, the power of the mind can cause objects to move. In telepathy, mind events influence the events in another person's mind. The effect of a specific mind event occurs, paradoxically, at more than one location. One location in space and time is, of course, within the flow of mind events that constitutes a person's consciously experienced identity. It is anchored in the physical structures of the brain. Infinitely many other locations, however, exist as the *possibility* of identical influences throughout space and time. When and where, and to what extent, these influences cause mental or physical events depends on the above-mentioned *resonance* at points of encounter between the wave patterns of the expanding spherical surfaces.

The Physics of Encounter suggests that a point of encounter is one of the endpoints of an oscillating string. The two endpoints of the string are located opposite each other in the surfaces of two QV-bubbles just before an encounter between the two expanding surfaces. An encounter between the surfaces of the two bubbles is an event that connects two points, one each in the surfaces of the smallest possible bubbles. Surfaces, as I said, are interconnected points.

In terms of high school geometry, we could say that two points that are opposite each other in two surfaces expanding toward a point of encounter will merge and become one point when the encounter occurs. But in the elementary reality of the proposed scenario, there are no visible lines and points, no surfaces that move through space. There is only a hidden, non-measurable energy that oscillates between two points: the energy of oscillating strings.

This energy causes the expansion of QV-bubbles. To visualize the process, think of the energy that instantly flows from one point to the other as a *line of thrust* that is like a bolt of lightning. Think of what you see when you watch a fireworks display. Singling out one point in the dazzling spectacle, you see the point explode into a multitude of lengthening bright lines. Their endpoints are like the outline of a luminous sphere. The bright lines are lines of thrust and their endpoints are the surface of the sphere. The sphere ceases to exist as the bright lines fade away when their energy is spent. To assure the continued existence of the sphere, at least for a while, a multitude of events would have to occur at the same time, one at each endpoint of the lines of thrust. Each event would again have to be an explosion, but a smaller one this time, again sending lines of thrust into all directions. The surface of the luminous sphere would then appear to consist of luminous bubbles.

If you imagine the same type of events in the surfaces of these bubbles, in a continuing sequence of fireworks explosions in the surfaces of the bubbles, you see an expanding spherical array of *increasingly smaller bubbles*. The bubbles on the inside of the spherical array disappear when the energy of the thrusts is spent, but additional bubbles continue to appear in the expanding surface of the sphere. Their number increases as the expansion continues.

This process will, of course, soon run out of steam. As the bubbles get smaller and smaller, down to the smallest possible bubble, the expansion of the sphere would eventually come to a halt. What keeps the process going? Only a continuous infusion of energy can assure a continuous expansion. This energy is part of the process that creates time.

In the proposed scenario of this process, *countless* lines of thrust are directed outward, into all directions, from the point where a QV-bubble originates. Within the spherical configuration of their endpoints (i.e. within the expanding surface of a sphere), the separation between any two endpoints located next to each other is not a separation in *space*, but a separation in *time*. The "length" in the dimension of time is very short compared to the corresponding length of the thrust in the dimension of space.

The geometry of the outward thrusts in the three dimensions of space can be compared to the two-dimensional geometry of the spokes in a bicycle wheel. The distance between any two endpoints of the spokes located next to each other in the rim is very short compared to the length of the spokes. Even if there were an infinite number of spokes, there would still be separation between their endpoints at the rim because the spokes are not parallel, but extend away from the center of the wheel. This geometric relationship is equally true for the countless thrusts within the three dimensions of space that are directed outward from a central point.

Together, their endpoints are a spherical configuration. The following pages will show why the expansion and collapse of this spherical surface corresponds to the process that creates time.

The bubbles in the spherical configuration of bubbles interact with one another. They expand and collapse where their surfaces encounter one another. Each point of encounter is the origin of a "fireworks explosion". The point is the endpoint of an oscillating string toward which the energy of countless other thrusts is directed in a process that involves a "resonance" of interacting, unobservable energy waves. The surfaces of all QV-bubbles are waves consisting of smaller QV-bubbles, expanding and collapsing. The smallest "something" in this process is larger than a point but not yet a bubble, as described above. It is an oscillating string.

Since each bubble expands into all directions, the process that creates an expanding bubble is not only like an explosion, but also like an *implosion*. When bubbles within the spherical configuration of bubbles expand, half of the surface of each bubble expands outward, *away from* the point where the spherical configuration (the "fireworks display") originated, but the other half expands inward, *toward* that point. An implosion, as interpreted here, is like an inward expansion. As part of a process that begins on a spherical surface and impacts on the origin of that surface, an *inward expansion* creates the inside reality of our consciousness, and with it the elapse of time as we experience it. An *outward* expansion creates the elapse of time outside our mind, in physical reality.

All observable physical events are like the outward expansion of a fireworks display. Implosions are unobservable inner events. The proposed scenario shows why both processes are interrelated and why they do not "run out of steam". The surface of the expanding spherical configuration, as I said, is a wave. When that surface is encountered by a wave that resonates with it, the surface collapses because the effects of the implosion cancel the effects that were created at the point where the expanding bubble originated. When a bubble "collapses", it disappears from the three dimensions of space *-relative to the observer*. For the observer, it is now one of the countless bubbles in a larger spherical configuration that is a different dimension - the dimension of time. As these bubbles interact, the process of expansion continues. The larger spherical configuration, or bubble, in this scenario was, in turn, created at a point in the expanding surface of an even larger bubble. All such points are points of encounter between resonating wave patterns.

The process of expansion does not, of course, continue indefinitely. It must be understood in terms of the influence of consciousness on physical events, and vice versa. The mind of the observer is an integral part of the above process. Successively experienced moments in time are created as the wave patterns of QV-bubbles interact. *The observer is part of the wave* at the location where the wave passes through a point in space. The observer contributes to the creation of the wave at that location. More about that later. What needs to be clarified first is how the thrusts of oscillating strings create a reality that is different from the dimensions of space: the reality of time. To describe that process, I will have to switch metaphors.

Think of a line of thrust as an ocean current that flows along the shoreline of a beach. If two currents of equal strength flow into opposite directions along the shoreline, they create a spot that is dangerous for swimmers. At the point where the two currents encounter each other, their effects cancel each other along the shoreline and create a current that flows away from the shoreline. Swimmers at that location are swept out to sea.

What happens at the point of head-on encounter between the two ocean currents illustrates the cancellation of lines of thrust within one dimension of space (the shoreline) and their transformation into a thrust that occurs along a *different dimension*. It is a line of thrust that runs at a right angle to the shoreline. In the metaphorical example of the shoreline, the thrusts are deflected into only one direction. The shoreline blocks the other direction. In the invisible physical reality of oscillating strings, thrusts toward each other are transformed into thrusts away from each other, into both directions and in continuous alternation.

This process corresponds to what could be called the *absorption* of equal and opposite effects, at a point of encounter, into a new reality created at that point. The reality is radically different because of the right angle that distinguishes each new event from the event that created it. In the unobservable, elementary reality of oscillating strings, the "absorption" at a point of encounter can deflect lines of thrust into a dimension that is not one of the three dimensions of space. This is what happens when countless thrusts converge on a common point of encounter from all directions of space. They create the dimension of time.

The thrusts, in this case, originate at a spherical array of points located around the common point of encounter. All thrusts collide head-on at that point. The thrusts cancel one another in all dimensions of space and are deflected into a new dimension. The deflection is like a rotation of all thrusts by 90 degrees. The new thrusts occur at right angles to the cancelled thrusts. The new thrusts, therefore, are "rotated" out of all three dimensions of space. Geometrically, the new reality created by this event is a spherical configuration consisting of interrelated points. It is like a surface that surrounds the point of encounter. The surface is a "fourth dimension" because all three dimensions of space intersect the spherical surface at right angles. The surface is not located in space because the thrusts have cancelled one another in all dimensions of space. The thrusts cannot, logically, rebound into the space in which they have cancelled one another. They are deflected, or absorbed, into a different dimension, just like the thrusts that I compared to ocean currents in the metaphorical image described above.

The creation of a *surface* by equal and opposite *lines* of thrust can be visualized by imagining two garden hoses squirting thin streams of water at each other. If the separation of the nozzles is minimal, and if the nozzles are pointing directly at each other, the water will splash outward in a plane. The surface of the plane corresponds to what string theorists have called a "brane". (The term is derived from the concept of a membrane. More about that in later chapters.)

The spherical array of points created by the encounters described above is like a surface that encloses the quantum vacuum of a QV-bubble. The thrusts of oscillating strings inject vacuum energy into the bubble. The two endpoints of a string are locations where vacuum energy is absorbed or released, in continuous alternation. Each endpoint is shared by countless other strings, or lines of thrust. The spherical surface is the dimension of time. It is a spherical array of interrelated points just like the metaphorical fireworks explosion.

The event *creates time* by creating a surface that does not expand into the dimensions of space. Time elapses as the expanding spherical surfaces interact at points of encounter. As described in the preceding chapter, the surfaces interact only if both surfaces *resonate* with each other. If the interaction occurs, the surfaces collapse and new QV-bubbles expand from the points of encounter. The spherical surface consisting of interrelated points is a *closed surface* because the points are connected by oscillating strings. The oscillations between the endpoints of the strings occur *within* the surface. The surface expands as additional QV-bubbles are created within it.

The interaction of expanding spherical surfaces consisting of QV-bubbles creates the quantum vacuum fluctuations of space. When the surfaces and the QV-bubbles within them collapse at points of encounter, the new QV-bubbles created at these points once again expand into spherical surfaces, which continue to expand until they are encountered. The size of these spheres, therefore, can vary. The surfaces of the spheres may expand across long distances within the universe, or interact here on earth. Since the surfaces of these spheres contain the vacuum energy of QV-bubbles, I will call them *VE-spheres*.

The creation of time through the interaction of the QV-bubbles in the surfaces of VE-spheres occurs when this interaction creates elementary matter events and elementary mind events. Time is a process. "There is more to time than just a dimension". The physicist Avshalom Elitzur made that point at a scientific conference on the "Frontiers of Time".(1) His reasoning supports the scenario proposed here. It has far-reaching implications. Since expanding and collapsing QV-bubbles in the surfaces of VE-spheres are the quantum vacuum fluctuations that pervade all of space, we can assume that these fluctuations are a source of hidden energy in the universe. The energy is hidden because it does not exist in the three dimensions of space. It is the energy that *creates* time. In other words, our universe consists of two kinds of reality. The quantum vacuum fluctuations of space are what the physicist Hal Puthoff called a *pre-manifest* reality. The reality we observe and experience is *emergent* reality.

The hidden energy is *vacuum energy*. The fluctuations of this energy occur at points of encounter between QV-bubbles in the expanding surfaces of VE-spheres, and these hidden events create time by causing observable events in space. Each point of encounter exists in three-dimensional space like the tip of an iceberg. In this scenario, the tip of the metaphorical iceberg is a zero-dimensional point. These unobservable points in space are the sources of hidden energy. This is why, as Puthoff said, "space is a player, not just a stage".(2)

The metaphor that "space is a player" corresponds to Einstein's reasoning that the "geometry of space" causes objects to move, relative to one another, in accordance with the gravitational attraction between them. As I mentioned in chapter one, the influence of gravity is not a measurable physical force like the electromagnetic radiation of light. The influence of gravity can only be *calculated* from the mass of the moving objects and from the way they move. Einstein used the term "point-objects" to describe how the effects of gravity can be mathematically pinpointed in the geometry of space. As interpreted here, these point-objects are points of encounter between QV-bubbles in the quantum vacuum fluctuations of space. In the proposed scenario, the effect of gravity corresponds to the minimal amount of vacuum energy in the "tip of an iceberg" that is hidden in a sea of fluctuating vacuum energy.

In this scenario, the anomalous events that seem to defy the known laws of physics are energized by abnormally large amounts of vacuum energy. The scenario describes the relationship between gravitational effects, vacuum energy, and the "dimension" of time. For simplicity, I will continue to use the word dimension in this context, even though, as Elitzur said, "there is more to time than just a dimension".

In the pre-manifest reality of interacting QV-bubbles, time does not (yet) exist because at that level of reality, there are no measurable matter events. Matter events are created by interactions between elementary particles. As explained, each particle is a "quantum", or certain amount, of measurable energy. In the theory of interactions between particles ("quantum mechanics"), such interactions create the "quantum systems" of observable matter. Where these quantum systems do not exist, the process that creates time does not occur. As Briggs explained (1/2), "time does not enter the picture until a quantum system interacts with its surroundings."

>A reminder: the numbers in parenthesis and separated by a slash identify the number of the chapter and the number of the reference note, respectively, where a quote was first mentioned and elaborated.<

The arguments of Elitzur, Puthoff, and Briggs support the scenario proposed here. When expanding QV-bubbles interact, this invisible, timeless process makes space a "player" in creating the reality we observe. The interaction takes place at points of encounter between the expanding surfaces of VE-spheres, because these surfaces consist of QV-bubbles. In visualizing this scenario, the distinction between a QV-bubble and a VE-sphere should be kept in mind. A QV-bubble is a quantum vacuum, but the space enclosed by a VE-sphere is not. The vacuum energy of QV-bubbles exists in the instantly expanding *surfaces* of VE-spheres.

For simplicity, I will occasionally use the term *VE-surface* to describe the effect of the vacuum energy that exists in the surface of a VE-sphere. The effect creates an elementary event (a gravity effect) at a point of encounter. The oscillating strings that inject vacuum energy into QV-bubbles connect the timeless, hidden reality of quantum vacuum fluctuations with the reality of observable events. One endpoint of a string absorbs the vacuum energy in the hidden reality, the other endpoint releases that energy into the reality of time and space as we experience it, in instantaneous alternation. Through the oscillating energy within the metaphorical "strings", the two realities become inextricably intertwined.

When countless oscillating strings converge on a common point of encounter from all directions, that point becomes the origin of a QV-bubble. The Physics of Encounter suggests that the location where a QV-bubble begins to expand corresponds to what the mathematician and computer scientist Michael Manthey called an *initial void*. Such locations, he said, are always either "empty" or "full", which corresponds to the digits "zero" and "one" in the "either/or" (binary) process of digital computation and information transfer. In this context, we could say that the two endpoints of oscillating strings are located, in alternation, in two intertwined realities of which one is "empty", filled only with vacuum energy, and the other is "full", in the sense of containing the energy of the interacting particles that are the building blocks of matter. Manthey called the locations of such binary processes *event buffers* where "we can get something from nothing".(3) As interpreted here, these event buffers are QV-bubbles that expand from "nothing" (a quantum vacuum) and create "something" (particles) at interrelated points of encounter. The process that creates particles from interacting QV-bubbles will be described in later chapters.

The philosopher Slavoj Zizek used an image that is strikingly similar to Manthey's concept of an initial void. He argued that not only matter, but consciousness, too, arises from "nothing" in terms of observable physical space. Consciousness, he said, "explodes" from a "non-All, a gap, a hole, in reality itself, filled in by phenomenal experience. (...Consciousness...) arises from an irreducible external *Anstoss* (...i.e. disturbance...). The subject emerges through the disturbance."(4)

In Zizek's scenario, which also supports the metaphorical images proposed here, the word "subject" stands for the subjective reality of consciousness, which enables persons to act on their own accord instead of being a passive object. The German word "Anstoss" can be translated as a push, or thrust. As interpreted by the Physics of Encounter, it is the thrust of an oscillating string that occurs in the surface of an expanding QV-bubble at a point of encounter with another QV-bubble. When the expanding surfaces of two bubbles interact, a new bubble is created at the point of encounter.

The "disturbance" is the collapse of the interacting QV-bubbles. The "explosion" of consciousness described by Zizek corresponds to the instantaneous, timeless expansion of a new QV-bubble. The word "phenomenal experience" stands for the conscious mind experiencing the phenomena created by physical processes.

A phenomenon, according to the dictionary, is "any event or circumstance that is apparent to the senses", and secondly, "in Kantian philosophy, a thing as it appears in perception as distinguished from the thing as it is in itself, independent of sense experience." I already mentioned the German philosopher Immanuel Kant who argued that the objective reality outside the mind, "the thing-in-itself", is unknowable. I mentioned that the atoms of all "things" are mostly empty space containing invisible forces, and that the atoms of a barking dog can influence the atoms of your brain, thereby triggering the creation of something that was not there before: your consciousness, which did not exist while you were sound asleep.

The Physics of Encounter shows that, at the elementary level of reality, consciousness and the physical reality that creates it are two sides of the same coin. Egon Freiherr von Eickstedt, a German physicist, described the relationship in terms that correspond to the scenario proposed here. Elementary reality, he said, is a "a focal point of density and at the same time the expansion of that point into a field. That reality is not a thing, but an event. The self is created as the outward flow of such events meets the oncoming flow of non-self events."(5)

The "self", of course, is the conscious mind of a particular person. "Non-self events" are physical events that occur outside that person's mind. The "focal point of density" is a point of encounter between two QV-bubbles that are located, respectively, in the surfaces of two interacting VE-spheres. The density is created by the countless thrusts of oscillating strings that exist in an expanding spherical surface. The "expansion of that point into a field" is the expansion of the new QV-bubble that is created at the point of encounter. The "outward flow" is the expanding spherical surface containing the "self". The "oncoming flow" is the expanding spherical surface that impacts on the mind of the observer of physical reality.

Wolfgang Pauli, an Austrian physicist who won a Nobel prize for his work on quantum mechanics, used a similar image for his description of the interaction between consciousness (the "psyche") and physical reality: "From an inner center, the psyche seems to move outward, in the sense of an extraversion, into the physical world."(6) The word "extraversion" can be equated with the concept of expansion. The "inner center" corresponds to the point of origin of an expanding QV-bubble. It is a point encounter between QV-bubbles. The encounter and the resulting creation of a new QV-bubble is an elementary event. It is an encounter between the mind and the physical reality of events outside the mind.

In keeping with all of the above-mentioned reasoning by highly respected theoreticians, the process of mind/matter interaction described by the Physics of Encounter is based on the assumption that the properties of consciousness are created in interacting QV-bubbles. These properties, or characteristics, of consciousness are like the elementary "ingredients" of the reality that is experienced when the mind encounters the reality outside itself. These ingredients are not, of course, physical substances. To think of them as the *qualities of experienced reality* comes close to the dictionary definition of the appropriate concept. The word used by scientists is *qualia*. The singular form of that word is *quale*. It is pronounced qua-lee. The syllable qua is pronounced as in quality (alternatively, as in quail). Qualia, the dictionary tells us, are "independent, universal qualities such as whiteness and loudness". They are information supplied by the senses, and feelings, which have their "own particular quality without meaning or external reference".

The dictionary phrase "without meaning or external reference" is of special importance here. Think of the two realities that exist if a barking dog wakes you up. The first moment of your newly activated consciousness is an event "without meaning or external reference". You become aware of yourself. Before you open your eyes and start thinking, there is just an initially indefinable instant of "loudness" in your mind. At the same time, as I said, you might feel irritated about the termination of a refreshing sleep. The initially indefinable negative feeling and the loudness are qualia.

The same applies to the above-mentioned "whiteness". Try to imagine that your bodiless "self" is floating at the center of a sphere, and that the inside of the spherical surface is white. You would be surrounded by meaningless whiteness without any "external reference". You would not even know that the whiteness is a surface. If the spherical surface that encloses your moment of consciousness expands away from you, there is no way for you to know that this is what's happening. For you, nothing would be happening as long as the surface remains white.

Events begin to "happen" for you when the whiteness is replaced by various colors, and when other sensations and feelings enter into the mix of qualia. What creates these ingredients of consciousness? The Physics of Encounter suggests that qualia are created inside QV-bubbles by the vacuum energy of oscillating strings.

Theoretical physicists describe a string as a one-dimensional (!) reality, defined by its two endpoints. This means that the string itself has no "thickness". It is an invisible, non-measurable burst of energy. The string is as short as it can possibly be. It is so short that it is below the threshold of observation. What is an "observation"? When you observe the dog that awakes you, as described above, you experience the qualia that are created in your mind. The image of the dog in your mind is like a mosaic of qualia. What you experience as a pattern in space is actually a sequence of mental events that are the focal points of (subconscious) attention. These focal points are points in time. They expand into the experienced moments that are the qualia in QV-bubbles. Each moment fills the entire "space" of your mind. You do not realize it, but what seems like an instantaneous observation is like a scan of the points in time that you see as a spatial pattern.

The sequence of all-encompassing, mind-filling events creates the experience of space. Each event is an experience of uniform quality like whiteness or loudness, but these qualities change from one point in time to the next. When QV-bubbles interact, new QV-bubbles are created at points of encounter. They are points in time. Each new QV-bubble contains the changing qualia of consciousness.

When oscillating strings create qualia they also create physical reality. The string theorist Brian Greene pointed out that strings exhibit a wide variety of "vibrational patterns" and that these vibrational patterns correspond to the properties of elementary particles. He argued that "what appear to be (...) particles are actually different 'notes' on a fundamental string".(7) Greene's acoustic metaphor can be readily applied to all elementary ingredients of consciousness. The sound of "notes" in the mental "space" filled with qualia are vibrations in the quantum vacuum of QV-bubbles. Strings create expanding QV-bubbles and oscillate within them. The vibrations are, of course, a metaphorical image, not like the vibrations of sound waves or the electromagnetic vibrations of light waves. But sounds, colors, and all other qualia are the "given" properties of a conscious mind and in that sense are like the equally abstract characteristics of a particle such as its electric charge, its mass, or its spin. As I explained, to say that a particle has "spin" does not mean that "something" actually rotates. Physicists have deduced the existence of spin from the changing characteristics of what is assumed to be the surface of a particle.

The qualia in QV-bubbles allow a conscious mind to experience events in space and time. The expansion of QV-bubbles and their interaction at points of encounter, however, are timeless events in a *pre-manifest* reality. The building blocks of our everyday experience, on the other hand, are an *emergent* reality that consists of elementary particles. Particles are created at interrelated points of encounter between QV-bubbles. The interrelated points correspond to what physicists call a *coupled map lattice*. A lattice, in this context, is a three-dimensional pattern of points. It couples, or connects, mind events created by the brain with events (interacting QV-bubbles) outside the brain. The Physics of Encounter describes this process. It is *dialectical* process in which time and space are created by timeless events.

A dialectical process is one that is best described in terms of what seems to be a logical contradiction. How can time be created by timeless events? Dialectical processes resolve the paradoxes of elementary physical reality. Philosophers who espouse the concept of dialectics regard it as a dynamic principle inherent in nature. It states that every elementary effect generates an opposite effect, and that the two opposites energize a process that makes them part of a larger whole. In that sense, the effects at the two endpoints of oscillating strings could be called dialectical opposites. The strings create and connect two intertwined levels of reality: the hidden level of reality consisting of quantum vacuum fluctuations and the observable level of emergent reality consisting of quantum systems. The vacuum and the non-vacuum are part of a larger whole that encompasses opposites. They are what Zizek called an "All", meaning: the totality of our universe. In our universe, the elementary ingredient of consciousness is the quantum vacuum in a QV-bubble. It is, as Zizek said (3/4), a "non-All, a hole in reality" that is "filled with phenomenal experience", i.e. with consciousness.

The counter-intuitive aspects of the elementary, hidden reality cannot always be adequately expressed in our everyday language. For simplicity, I will use the customary grammatical structures. Describing the timeless interaction between QV-bubbles with reference to a specific outcome as an "instantaneous succession of events" is a case in point. (Language buffs will readily identify that phrase as an oxymoron.) The logic of language distinguishes what happens now from the future and the past. The expansion of a spherical surface containing QV-bubbles is, logically, the cause of a future event. The event will not occur until the expanding surface is encountered. But the expansion of the unobservable surface is instantaneous (a timeless event), therefore it is not necessary to "wait" until the encounter occurs. The future is contained in the expanding surface as a broad spectrum of possibilities, even though the encounter has "not yet" occurred. I will use sentences that make the customary distinction between past, present, and future, even though the hidden reality in our universe exists in an eternal NOW.

Occasionally, during an *altered state of consciousness*, we may catch a glimpse of the interconnectedness of all reality that exists where "time does not enter the picture". The most striking and awe-inspiring example of this kind of event is a *Near-Death-Experience* (NDE). Thanks to the advances in medical science, an increasing number of people have been revived from clinical death. Their heart had stopped beating. Their brain was no longer functioning. Many had an NDE. Their conscious mind had hovered above their lifeless body and had watched what was going on. Then it left the scene and went on a journey into another world. When it "returned" into the body that was restored to life, it was an enlightened mind. Through an NDE, people who were brought back to life acquired an unshakable confidence that the mind survives the physical death of the body.

In the metaphorical scenario proposed by the Physics of Encounter, the human mind "travels" with the expanding surfaces of VE-spheres because the surfaces consist of the QV-bubbles that contain the basic ingredients of consciousness. Within the world of quantum events, which is the world that includes the atoms in the brain, the human mind does not stray very far from the biological reality of the brain in which it is rooted. Things are radically different when the mind enters the uncharted realm of timeless events during altered states of consciousness. These occur not only during NDEs, but include encounters with UFOs and "aliens".

In the hidden reality of fluctuating vacuum energy, encounters between QV-bubbles containing the qualia of consciousness occur if the expanding VE-surfaces containing the metaphorical bubbles are "in resonance". The concept of resonance will be explained later. Without this resonance, no encounter occurs. The VE-surfaces pass through one another and continue to expand. As interpreted here, encounters correspond to what physicists have called the *potentialities* (possible events) of mind/matter interaction. Robert Jahn described these potentialities as "ineffable". The dictionary defines ineffable as "too overwhelming to be expressed in words, as in *ineffable beauty*, or too awesome or sacred to be spoken, as in *God's ineffable name*".

Jahn has emphasized that a description of the physical processes underlying the reality of consciousness requires "the generous utilization of conceptual metaphors" to access a "sea of ineffable, complexly intertwined potentialities".**(8)** This endeavor is well worth the effort. If we want to understand consciousness, we must understand what allows us to see beauty and fills us with awe. We must understand the events that cannot be explained by the currently known laws of physics. They are caused by the hidden energy in a timeless reality. A daunting task - and an exciting challenge.

> *It is impossible to meditate on time*
> *and the mystery of the creative passage of nature*
> *without an overwhelming emotion*
> *at the limitation of human intelligence.*
> Alfred North Whitehead

Based on the assumptions presented in this chapter, here is what the Physics of Encounter says about the processes underlying consciousness and the perplexing aspects of time.

1) Time is more than a dimension. Like space, which is "a player, not just a stage", as Puthoff said (3/2), time is a process that involves the observer.

2) The process occurs where the self observes (interacts with) non-self reality. It occurs when mind encounters matter. (In this context, the terms "self" and "observer" have the same meaning. "Non-self" reality, as defined below, consists of unobservable processes underlying the emergence of matter and of other minds.)

3) The location of the observer, as defined here, is the location of the QV-bubble that contains the qualia (the basic ingredients of consciousness) experienced by the observer at the instant the QV-bubble interacts with non-self reality. QV-bubbles exist in the expanding surfaces of VE-spheres. All QV-bubbles in a VE-surface are alike. Non-self reality, therefore, exists outside that surface. It exists in the surfaces of other VE-spheres. (The distinction is complex because every surface is a pattern of superposed, intertwined surfaces. More about that in later chapters.)

4) Time is the dynamic boundary, or horizon, of the observer's experienced moment. It is the expanding surface of a VE-sphere. The expansion (the flow of time) is a continuous, smooth, uninterrupted process. As soon as an encounter occurs, a new VE-sphere instantly expands from the point of encounter on the expanding surface.

5) Nothing actually moves. The invisible boundary of the self is continuously re-created as new VE-spheres expand from the point in time where the self is located. All points on the expanding surface of a VE-sphere are equivalent points in time.

6) As soon as the surface of a VE-sphere that contains the observer's present moment has encountered the non-self reality of another VE-sphere, both surfaces cease to exist for the observer. This corresponds to what physicists have called the collapse of a probability wave. As interpreted here, the wave is the surface of a VE-sphere expanding toward the observer. The wave ceases to exist for the observer because the encounter creates a new experienced moment for the observer. It is a new QV-bubble that expands into a new VE-sphere.

7) Nothing actually collapses or disappears. The thrusts of oscillating strings converging on common points of encounter continuously create the surfaces of QV-bubbles and VE-spheres, without interruption. They are the quantum vacuum fluctuations that create the emergent reality of mind and matter, of consciousness and time. Specific configurations of encounters contribute to the creation of specific qualia and to the quantized effects of elementary particles.

8) The collapse of a probability wave corresponds to the "absorption" of equal and opposite thrusts into a point of encounter. I described the process in this chapter by comparing it to the opposite thrusts of ocean currents along a shoreline. The encounter of the thrusts creates an additional dimension, i.e. right-angle thrusts directed away from the original flow and away from each other.

9) The opposite thrusts within the additional dimension, directed away from the point of encounter, correspond to the two directions of time that exist at the elementary, non-observable level of reality. Relative to the present moment experienced by the observer, therefore, the effect of the encounter can cause an event in the future or in the past.

10) The two possible effects of an encounter, in the future or in the past, co-exist as equal possibilities in a probability wave, i.e. in the expanding surface of a VE-sphere. The two possible effects can be visualized as the two hemispheres of an expanding VE-sphere that is created at a certain distance from the observer. One hemisphere expands toward the observer and the other hemisphere expands away (recedes) from the observer. Relative to the observer, the oncoming hemisphere expands forward in time and the receding hemisphere expands backward in time. When either hemisphere is encountered at any one point, the entire wave (the whole VE-sphere) collapses. Therefore, when one of the two possible effects creates an event, this pre-empts the other effect.

11) Time only "flows forward" for an observer whose consciousness is rooted in a brain because the stable configurations of encounters in the atoms of the brain limit the self to encounters with what von Eickstedt (3/5) called "the oncoming flow of non-self events". The observer's body and brain remain at a fixed location while the invisible waves created by each act of observation (an encounter involving the atoms of the brain) expand away from the point of encounter.

12) Mind events mediated by the brain create effects that correspond to what physicists call retrocausation. They cause events that occur in the observer's past. The retrocausation effects are transmitted by the expanding surfaces of VE-spheres created at points of encounter involving the atoms of the brain. Since they are surfaces receding from the points of encounter, they transmit effects backward in time, relative to the observer's present moment.

The effects described above are possible effects that will not necessarily cause observable physical events. They correspond to what physicists call potential energy. They are the invisible (pre-manifest) reality of vacuum energy. The expanding surfaces of VE-spheres do not travel through observable space. The instantaneous expansion occurs in the dimension of time. It creates time as a continuous sequence of encounters involving the observer.

## NOTES

(1) Elitzur (2006), conference presentation
(2) Puthoff (2001), conference presentation
(3) Manthey (1997), p. 14
(4) Zizek (2006), pp. 197 and 223
(5) von Eickstedt (1954), p. 7
(6) Jung and Pauli (1955), p. 175
(7) Greene (2000), pp. 145-146
(8) Jahn and Dunne (2004), pp. 557 and 560

## Chapter 4

**Atomic nuclei and the inner light of consciousness. "The spark that ignited the big bang." Black holes, implosions and retrocausation. Lederman's God Particle.**

The process that generates consciousness in brains is not an abstruse, indefinable reality. It is firmly rooted in the nitty-gritty reality of interacting atoms. The focal point of the process is in the nucleus of an atom. The nucleus is the dense core of an atom, an unimaginably high concentration of matter and energy. Einstein showed that matter and energy are interchangeable aspects of physical reality.

In an atomic nucleus, as interpreted by the Physics of Encounter, the two manifestations of physical reality, matter and measurable energy, constantly fluctuate between being either one or the other. The medium in which these fluctuations occur is non-observable vacuum energy. In other words, the core of an atom is the site of extremely intense quantum vacuum fluctuations. Physicists have described it as a fiercely hot *plasma*. It is like a violently agitated, boiling and bubbling substance.

Quantum vacuum fluctuations are like invisible bubbles that expand into non-measurable, invisible waves. The waves instantly reach all possible locations in observable space by expanding in the unobservable substratum of space. The waves interact, expanding and collapsing at points of encounter, if the requisite resonance exists at those locations. I will explain the details of the resonance process a little later.

At points of encounter, the waves are absorbed into atoms, agitate the electrons in those atoms, and wind up as high-intensity encounters in atomic nuclei, where the cycle of expansion and encounters begins anew.

In the scenario proposed here, particles are created at interrelated points of encounter between the above-mentioned waves. The waves, as I explained, are expanding VE-surfaces. The unpredictable encounters, resulting from the resonance between interacting surfaces, cause particles to pop into existence, or to disappear just as unexpectedly. Particles may suddenly change their characteristics, or decay into less massive particles. All of these events are caused by encounters between VE-surfaces. Whether particles appear or disappear, suddenly acquire or lose energy in so-called quantum jumps, or disintegrate, depends on the direction into which the VE-surfaces expand, relative to the observer. The expansion may be an explosion or an implosion, as described in chapter three. If the expansion is an *explosion*, the energy creates observable physical reality at interrelated points of encounter, i.e. the mass and the energy of elementary particles. If the direction of the expansion is away from the observer, the event is an *implosion* that takes away mass and energy.

The Physics of Encounter suggests that in the world of normal particle interaction, an implosion creates a miniature *black hole* that instantly disappears again. The metaphorical hole is like a vacuum cleaner that sucks in the mass and the energy of particles. The hole is "black" because it absorbs the energy of light. The specter of powerful and possibly dangerous black holes created by an ambitious physical experiment was very much in the news in 2008 and again in 2009. The experiment at the nuclear research laboratory CERN outside Geneva was designed to smash subatomic particles against each other in a so-called collider, or particle accelerator, at extremely high energies. The purpose was to shed some light on the controversial issue of the proverbial "big bang" and to identify the very first elementary particles that were thrust into existence when that event occurred, presumably some 15 billion years ago.

The more fundamental question, of course, is whether the big bang was an explosion that created our universe "out of nothing". Did our universe need an "act of creation"? Is it conceivable that it is an eternal reality, with no beginning and no end? Could the cycles of expansion and collapse that characterize the unobservable reality underlying our observable world be mirror images of cosmic explosions and implosions?

This issue touches upon the nature of consciousness and the ultimate source of reality. It is, therefore, also a religious issue. The model of mind/matter interaction proposed in this book shows that science offers ample evidence in support of the assumption that consciousness pervades an eternal, "immortal" universe. This allows scientists and religious thinkers alike to enrich the concept of God with additional, broadly acceptable meaning. The Physics of Encounter buttresses the argument that the self survives the death of the physical body and then shares the wisdom of a Universal Mind. Near-Death-Experiences and a broad spectrum of other anomalous events provide glimpses into a reality that lies beyond the world revealed to us by our senses.

The Physics of Encounter suggests that this is the world of vacuum energy. As interpreted here, the explosive encounters of particles in colliders like the one built for the CERN experiment tap into that energy. That experiment was designed to simulate the conditions that presumably existed a split second after the big bang. By studying the particle debris created by the collisions, physicists said, it might finally be possible to verify the "standard model" of the inner workings of the universe.

As reported in an editorial of the *Boston Globe*, published in the *International Herald Tribune* before the experiment began, "a few researchers fear that the collider will create black holes that will expand and suck in the entire world. (...) It would be strange and sad if, in the course of establishing how and why matter emerged out of nothingness, an experiment were to consume a planetful of people who might be interested in an answer."(1)

The experiment may have taken place by the time this book is printed. I will have more to say about it toward the end of this chapter. The experiment needs to be understood in terms of the scenario proposed here, which suggests that the big bang was an encounter between two immensely powerful VE-surfaces involving the entire vacuum energy in our present-day universe. Was the event an explosion from "nothing"? The point where the explosive encounter occurred corresponds to the "initial void" described by Manthey (3/3). In the elementary physical processes that occur all around us today, that "void" corresponds to the "hole" in observable physical reality described by Zizek (3/4). In Zizek's scenario, it is filled by "phenomenal experience", i.e. by the basic ingredients of consciousness.

The big bang was not just an explosion. It was also an implosion. It was an encounter between two realities, the cosmic equivalent of what constantly repeats itself in the quantum vacuum of atomic nuclei. The big bang created not only the universe of observable matter and measurable energy, but also the *other space* of invisible reality, of hidden energy, of phenomenal experience, which is intimately intertwined with the reality studied by physicists. That is the reality of consciousness. It pervades our universe. It arises from the quantum vacuum fluctuations of space. It is a cosmic consciousness that corresponds to the "sea of ineffable potentialities" described by Jahn and Dunne (3/8). The potentialities are the hidden energy in our universe.

Implosions in the world of particles and elementary reality, as interpreted here, create the paradoxical effect of causation *backward in time*. Physicists have established, through experiments, that the counter-intuitive phenomenon of causation backward in time does, indeed, exist and have called it retrocausation. The eminent physicist John Wheeler argued that retrocausation occurs on a cosmic scale and that we, through our consciousness, are engaged in retroactively creating our universe by observing it. This supports the scenario proposed by the Physics of Encounter. The cumulative effects of consciousness, of every act of observation, increase the amount of hidden energy in the universe. Some 15 billion years ago, the sum total of that accumulated effect triggered the big bang as it encountered the forward causation effect of what has become the measurable energy in our universe. The two unobservable effects are propagated in opposite directions in the dimension of time. They expand into all three dimensions of space, pervading our universe as forward causation originating in the "here and now" and as the backward causation that triggered the beginning of time. The process creates time as we are experiencing it at this very moment. The details of Wheeler's arguments, and of the conclusions we can draw from this scenario are presented in Part Two of this book.

The conclusions provide strong support for the argument that the broad spectrum of anomalous phenomena can be explained by taking a fresh look at the laws of physics. Claims by parapsychologists that it is sometimes possible, for example, to foresee the future, are in keeping with the evidence that retrocausation allows future events to affect events occurring at the present moment. Similarly, anomalous events occasionally allow us to experience the presence of a consciousness that transcends the limitations of time. Near-Death-Experiences and encounters with UFOs or aliens are the most frequently cited examples. The hidden energy of consciousness, or subconscious motivations, provides an explanation for the puzzling ability to make things move through the power of the mind (psychokinesis) and for the mind-boggling poltergeist events, which physicists have dubbed "recurrent spontaneous psychokinesis" (RSPK). The ability to heal other people through the power of the mind and of "bioenergy" also belongs into this category.

Last but not least, the insight that the big bang was, in essence, an encounter of the kind that continuously occurs in the nucleus of every atom in the universe, suggests that this cosmic event is part of a never-ending cycle of expansion and collapse and not a singular act of creation. The CERN experiment has sharpened the focus on the issue of black holes associated with the "creation of something from nothing", as Manthey put it.

In the model of mind/matter interaction proposed by the Physics of Encounter, black holes in the quantum vacuum of atomic nuclei absorb the light that impinges on the atoms of the brain and transforms it into what we could call the "inner light" of a conscious experience.

Research published by the biophysicist R. VanWijk and others suggests that photons travel not only through the open space between widely separated atoms, but are also exchanged between the interconnected atoms of brains and other biological matter. VanWijk argued that the "oscillation process" triggered by these photons has a "bio-informational aspect" and can be presumed to regulate cell division and cellular differentiation in accordance with the information that is absorbed into the nerve cells of the brain.(2)

The inner light of a conscious experience is the hidden energy in our universe. It is encapsulated in the vacuum energy of interacting QV-bubbles, which are the invisible "substance" of expanding VE-surfaces. QV-bubbles energize the expansion. The expansion occurs because the surfaces of QV-bubbles acquire QV-bubbles of their own. In the preceding chapter, I described that process by using the metaphorical image of a fireworks display.

A QV-bubble expands into a VE-sphere that eventually encounters "outside reality". That reality is a different VE-sphere expanding toward the observer. The *location of the observer*, in this context, is the location of the QV-bubble containing, at any one instant, the qualia of the observer's consciousness. The location is a point of encounter with a VE-surface that is part of what von Eickstedt (3/5) called the "oncoming flow of non-self events". This scenario connects the theory of interactions between elementary particles (quantum mechanics) with the relativity theory of space and time introduced by Einstein. Space and time, as interpreted by the Physics of Encounter, are continuously created, with the participation of the observer, by VE-surfaces that expand and collapse as they encounter one another. When an encounter causes these surfaces to collapse, a new QV-bubble instantly begins to expand from the point where the encounter occurred. The QV-bubbles created at points of encounter, one after the other *(relative to the observer)*, contain the changing qualia experienced as the flow of time by the observer of physical reality.

Points of encounter are points in time. More precisely, the points are located in the reality that the individual human mind experiences as time. They are located in the one-dimensional reality of oscillating strings. The unobservable oscillations occur between the two endpoints of the strings. Each of these two points is a point where an explosion and an implosion occur in alternation. Implosions create QV-bubbles that contain the changing qualia of the observer's consciousness. Explosions create QV-bubbles that expand toward a different observer. For the individual human mind, implosions create the experience of time, while the explosions carry effects through the space experienced by a different observer. Oscillating strings connect the dimensions of space and time.

Explosions create elementary particles. They are matter events. Implosions are mind events. Oscillating strings connect the realities of mind and matter. The two realities are "two sides of the same coin".

To understand this string scenario in terms of two intertwined realities, you should not visualize the string as a line stretched across the space you see in front of you. The string is like a line that stretches toward you, beginning at a single (invisible) point located in the space that surrounds you. The "string" is a thrust that creates the reality you experience. The point of impact is one of the endpoints of the oscillating string. The other endpoint is located "underneath" that point. Relative to you, the observer, the two points are *superposed* on each other. For you, they are like one point. All of the superposed *waves* in the surfaces of expanding VE-spheres consist of superposed *points*. Conversely, your mind events transmit undetectable thrusts into the space that surrounds you. Your mind contributes to the creation of the physical reality you observe.

How many superposed waves in the surfaces of VE-spheres collapse, in a cascade of encounters, depends on the degree of resonance at the points of encounter. The expanding surfaces correspond to the *probability waves* in quantum mechanical theories. The waves are expanding and collapsing QV-bubbles in the surfaces of VE-spheres. The waves superposed on one another in the surfaces of VE-spheres are what theoretical physicists have called *superpositions*. They determine the characteristics of the particles created at interrelated points of encounter.

When two VE-surfaces encounter each other and collapse, a new QV-bubble is created at the point of encounter. The QV-bubble expands into new VE-sphere. This does not mean that two VE-spheres are replaced by one new VE-sphere each time an encounter occurs. If that were the case, the process would consume its own "substance" and soon come to an end. The substance in the intertwined realities of our universe does not disappear. It oscillates between the observable *emergent* reality and the non-observable *pre-manifest* reality. The building blocks of observable reality are the quantum systems of interacting particles. The non-observable reality consists of interacting QV-bubbles that are the quantum vacuum fluctuations of space.

QV-bubbles are created by oscillating strings. When QV-bubbles collapse at points of encounter, they "splinter" into the strings that created them. The strings become the lines of thrust in the metaphorical fireworks display described in chapter three. They expand away from a common point of encounter. The strings once again create QV-bubbles when they converge on a common point of encounter, in an unending sequence of explosions and implosions.

This aspect of the proposed scenario is supported by the conclusion of theoretical physicists that a quantum vacuum is not really "empty". It is filled with what are called "virtual particles". These paradoxical particles are continuously created inside the vacuum but never become observable because they instantly destroy each other in violent interactions. A quantum vacuum seethes with enormous energy. As interpreted here, virtual particles are created by the same process that creates "real" particles, but virtual particles result from implosions, not explosions. The oscillating strings that create the implosions also create the qualia of consciousness inside the quantum vacuum of a QV-bubble. Consciousness exists in the quantum vacuum, but it does not exist as a measurable energy outside the vacuum.

As reported by the science journalist Walter Leonhard, the astrophysicist Edward Tryon used similar reasoning when he argued that our entire universe might be the cosmic equivalent of a quantum vacuum. Relative to a hypothetical "outside observer", the energy effect of such a universe would be zero, he said. Vacuum fluctuations of cosmic proportions could result in the creation and collapse of an entire universe without violating the laws of physics.(3)

Our universe is filled with vacuum bubbles of cosmic size. The Harvard-Smithsonian Center for Astrophysics reported that galaxies and other cosmic constellations seem to float on huge spherical surfaces. As described by the German newsmagazine *Der Spiegel*, the data suggest that the entire universe consists of "tightly packed, hollow spheres, jostling one another like soap bubbles in a foam bath." The "cosmic bubbles", as explained in the article, are invisible because they are empty space. Where they "touch one another", the concentration of stars and other cosmic matter is higher than elsewhere.**(4)**

As interpreted by the Physics of Encounter, the bubbles "jostling one another" in our universe are interacting QV-bubbles. Here on earth, QV-bubbles are small because they are encountered and collapse soon after they begin to expand. In outer space, where the density of matter is low, the probability of encounters is reduced. QV-bubbles that originate in outer space expand into a much larger size until an encounter finally causes their collapse.

The proposed scenario is supported by the fact that the concentration of cosmic matter is higher at locations where the invisible bubbles "touch one another". As explained above, the quantum systems that are the building blocks of matter are created at interrelated points of encounter between the expanding surfaces of QV-bubbles.

Evidence is accumulating that the universe in which we exist is just "one of many bubbles in a gigantic cosmic foam". That is how the science journalist Stefan Klein described the view of the astrophysicist Andrej Linde in the German newsmagazine *Der Spiegel*. Linde, according to Klein, argued that the energy of a fluctuating vacuum "triggers all physical events", including the proverbial big bang, the grand explosion that allegedly created our universe. Linde pointed out that, since energy fluctuations continuously occur throughout a vacuum, the big bang cannot be a singular event. Every time fluctuating vacuum energies "encounter one another in the right mix", he said, "a new universe emerges from that mix".**(5)** Encounters involving the "right mix" of energies correspond to what I have called the "resonance" of superposed wave patterns at points of encounter between the QV-bubbles contained in the expanding surfaces of VE-spheres.

The vacuum, as described by Tim Folger in the science magazine *Discover*, is a "strange region that spontaneously creates and destroys energy". Some astrophysical calculations, Folger explained, suggest that the present expansion of our universe is a "prelude to another stupendous infusion of energy from the vacuum in the far distant future. When that happens, the universe will essentially undergo another Big Bang." According to a theory based on these calculations, Folger said, the "sequence of Big Bang, expansion, Big Bang, would never end".**(6)**

The "Big Stream Theory" first proposed by the cosmologist Fred Hoyle also rejects the assumption that there was only one big bang. It states that the universe oscillates, expanding and collapsing in endless cycles. The estimated time period for each cycle, according to Hoyle, is between 40 and 100 billion years.

Most physicists cling to the original interpretation of the big bang theory, according to which the universe was created in a gigantic explosion some 15 billion years ago and will continue to expand because there is no known force to counteract the expansion. This scenario has been welcomed by many people as common ground for discussions about the relationship between science and religion. The physicist Halton Arp has criticized this point of view: "Let us examine for a moment the current all-encompassing science of cosmology, i.e. the physics of the universe. The big bang theory proclaims that the whole universe created itself instantly out of nothing. I believe there are many observations by now that disprove this, but even supposing for a moment it were true, would it be essentially different from the religious belief that God created the universe at some time in the past? (...) The Vatican has supported the big bang theory since they alertly sense a place for a 'creator'. (...) The most likely truth (is) probably something quite different from that of any of the current partisans."(7)

One aspect of Arp's argument does not do justice to the current theories of the big bang. Some of these theories acknowledge that the universe did not create itself "out of nothing". At issue is the nature of the highly compacted "something" from which the universe might have emerged. The Physics of Encounter suggests that it is what Michael Manthey (3/3) called an "event buffer" where, as he said, "we can get something from nothing". When the big bang occurred, the effects of two *virtual* events encountered each other and created one *real* event.

The encounter corresponds to what the German astrophysicist Günther Hasinger described as "the spark that ignited the big bang". That spark, Hasinger surmised, is the "dark energy that was only recently discovered in the universe, a force of repulsion that pushes the galaxies apart (...). We know virtually nothing about is source. (...) An answer to this question may shake the foundations of physics. There are many indications that we have come upon a hidden form of energy that exists in the seemingly empty vacuum."(8)

This hidden form of energy is the subject of this book. As interpreted here, it is contained in the QV-bubbles that expand into VE-spheres. The quantum vacuum in a QV-bubble contains virtual events, as described above.

Virtual events become real events (physical reality) when expanding VE-surfaces encounter each other. The "sea of (...) intertwined potentialities" in the metaphorical image used by Jahn and Dunne (3/8) is the vacuum energy released at interrelated points of encounter.

The reality inside a QV-bubble is an unobservable *virtual* reality for physicists measuring the effects of events that occur outside their own mind. But the events inside a QV-bubble are something *very real* when they are experienced directly, inside the mind. QV-bubbles contain the elementary ingredients of consciousness, the qualia of powerful emotions, for example, that are created when the mind encounters what von Eickstedt (3/5) called the "oncoming flow of non-self events".

When that "oncoming flow" is information gained from physical experiments, the joy that swells up after a scientific discovery is a rich source of energy. It triggers additional research, not only about the nature of our universe, but also about the nature of our consciousness. The Physics of Encounter shows how the two kinds of reality are intertwined.

> *The aim is not to degrade mind to matter,*
> *but to upgrade the properties of matter*
> *to account for mind, and to tell how*
> *from the dust and water of the earth,*
> *natural forces conjured a mental system*
> *capable of asking why it exists.*
> Nigel Calder

The computer scientist Michael Manthey coined the term *combinatorial bit bang* in his description of the relationship between the expansion of information and the expansion of the universe that occurred at the moment of the big bang. A "bit" is the technical term for an elementary piece of information. For a conscious mind, a "bit bang" is like the sudden presence of a new insight. The process, Manthey points out, does not result from "number crunching", but from a "synchronization" between events. The synchronization corresponds to what I have called the resonance between events. Manthey described the principle of the "combinatorial" creation of information in terms of a "coin demonstration":

"A man stands in front of you with both hands behind his back, whilst you have one hand extended in front of you, palm up. You see the man move one hand from behind his back and place a coin in your palm. He then removes the coin with his hand and moves it back behind his back. After a brief pause, he again moves his hand from behind his back, places what appears to be an identical coin in your palm, and removes it again in the same way. He then asks you, 'How many coins do I have?'. (...) The most inclusive answer to that question is 'One or more than one', an answer that exhausts the universe of possibilities given what you have seen, namely at least one coin."

That answer can be encoded in one bit of information regarding two possibilities: one or more than one. In Manthey's scenario, the man then extends his hand again and you see that there are two identical coins in it. You now know that there are two coins and that is another bit of information. Where did that bit of information come from? Well, obviously, from the fact that you saw two coins in his hand. When the coin demonstration began, you also saw two coins, first one, and then another. But since the coins were indistinguishable, seeing them one at a time did not tell you what you now know: the man has two coins. This new bit of information, Manthey explained, "originates in the simultaneous presence of the two coins". Manthey called this scenario a *co-occurrence*. "We see from the coin demonstration", he concluded, "that there is information (...) in the universe which in principle cannot be obtained sequentially."(9)

In this scenario, the important fact is that the coins are indistinguishable in every respect. They represent particles or geometric points. In the Physics of Encounter, co-occurrences correspond to the points of encounter in what physicists call *coupled map lattices*. As interpreted here, they include encounters that create both elementary matter events and elementary mind events, or acts of observation. A co-occurrence creates a quantifiable event, like the precise information about the number of coins. The encounters are interactions between the expanding surfaces of VE-spheres. The surfaces, expanding and collapsing at points of encounter, are invisible, non-quantifiable fluctuations of vacuum energy. The fluctuations become quantifiable events when they occur in a coupled map lattice, which includes the mind events of a human observer. The size of elementary particles in the dimensions of space and the duration of their existence in the dimension of time is determined by the continuously changing configurations of interrelated points.

The creation of particles and of mind events through the "bit bangs" that are encounters between the expanding surfaces of VE spheres mirrors the "big bang" that is said to have created our universe.

*Not once in the dim past,*
*but continuously, by conscious mind,*
*is the miracle of Creation wrought.*
Arthur Eddington

Physicists have discovered a steadily increasing number particles. To say that they have "discovered" the particles is a misleading way of putting it. Physicists have *created* most of them in particle accelerators. Known particles are accelerated in huge circular tunnels and put on a collision course. The collisions between the high-speed particles are like explosions. Bigger particles splinter into a large variety of smaller, high-energy particles.

Vexed physicists speak of a veritable "particle zoo" containing an overabundance of physical entities with split-second life spans. This makes it difficult to establish a suitably small number of simple criteria for classifying the particles. The Physics of Encounter provides such criteria, based on the scenario of interacting VE-surfaces that contain layers of QV-bubbles. The proposed classification is described in detail in Part Three of this book.

The process of quantifying, through measurements, the effects of particles is not as straightforward, however, as it may seem. The act of observing a physical event is inextricably intertwined with the observed event. That's because the *location of an observer*, as I explained, is the location of the QV-bubble that contains the qualia of the observer's consciousness at the moment it interacts with physical reality. Interacting QV-bubbles create physical reality. They also create consciousness.

QV-bubbles exist in the surfaces of VE-spheres. All QV-bubbles in any one of the layers of QV-bubbles in a VE-surface contain the same qualia. An *act of observation*, in this context, is an encounter between VE-surfaces in a coupled map lattice. From the definition of a coupled map lattice, it follows that one of the surfaces participating in the encounter contains the qualia of the observer's consciousness. The number of VE-surfaces involved in an act of observation depends on the type of particle that is created by the interrelated encounters.

It follows from the above scenario that the interaction between QV-bubbles results in what physicists call *universal constants*. These refer to the size of fundamental energy effects. If we visualize the energy of particles as the energy of bullets continuously fired from a machine gun, the energy of each particle is *always the same* if the machine gun continues to fire at the same rate. The firing rate is a metaphor for the wave aspect of energy. Waves have a *frequency*, which tells us how often wave crests and wave troughs alternate within a certain period of time. In the world of elementary particles, waves that oscillate at a higher frequency have more energy.

In the scenario proposed by the Physics of Encounter, a wave crest corresponds to the point where an encounter between expanding VE-surfaces is most likely to occur. The surfaces, as I explained, are layers of wave patterns (superpositions) consisting of expanding and collapsing QV-bubbles. Toward the outer layers, the QV-bubbles become smaller and smaller. The outermost layer consists of points that are just beginning to expand into QV-bubbles. Each point is a common point of encounter between the thrusts of oscillating strings. I described the process in the previous chapter by using the metaphor of a fireworks display.

Since VE-spheres interact only if there is a *resonance* between the wave patterns in their respective surfaces, they are always of precisely the same size when they encounter each other and collapse. An encounter occurs when a wave crest of one VE-surface fits into a wave trough of the other VE-surface. The surfaces *enmesh*. The process is like the enmeshment of two cogwheels, which rotate in unison and at the same speed if they are of the same size. They cannot enmesh if they are of different size and rotate at different speeds when they approach each other.

Instead of a continuous rotation, the enmeshment of VE-surfaces causes the cancellation of equal and opposite energy thrusts at points of encounter. I described the process in the previous chapter by using the metaphor of ocean currents encountering each other along a shoreline. The thrusts are the energy of oscillating strings. When the thrusts are cancelled in one dimension, they "disappear" in that dimension and "re-appear" in a different dimension. As explained, that is like the *absorption* of energy into a point of encounter. At that point, the thrust executes an abrupt 90-degree "rotation".

This absorption of thrusts at points of encounter causes the collapse of VE-surfaces. The energy of the thrusts is absorbed in a cascade of encounters between the QV-bubbles in these surfaces. Like the interacting VE-spheres, the QV-bubbles in the surfaces of these spheres are the same size when they interact.

This is why the QV-bubble that contains the observer's qualia of consciousness is always the same size as the QV-bubble encountered by the observer, even though the QV-bubbles in the quantum vacuum fluctuations of space have various sizes. The QV-bubble encountered by an observer is the physical reality that von Eickstedt (3/5) called the "oncoming flow of non-self events". The accumulated energy of the oscillating strings in the QV-bubble encountered by the observer corresponds to the universal constant that has come to be known as *Planck's constant*. It was named after the German physicist Max Planck. The constant speed of light, relative to an observer, is another universal constant. It can be similarly explained by the invariable size of the QV-bubbles involved in an encounter between the observer and the observed physical reality.

The mathematics used by physicists to define Planck's constant is based on the distinction between *waves* and *particles*. The two concepts seem to be incompatible. A wave is an uninterrupted motion of wave crests and wave troughs like the ones we see on the surface of a pond into which we have dropped a pebble. A particle, on the other hand, is like one of the machine gun bullets mentioned above. The bullets are fired one after the other, with interruptions. This distinction between uninterrupted and interrupted reality corresponds to the wave/particle paradox of physical reality. Physicists describe the distinction in terms of *continuity versus discontinuity*.

The thrusts of oscillating strings, as interpreted here, unite the wave aspect and the particle aspect of physical reality because both aspects are created by oscillating strings in the surfaces of VE-spheres. The wave aspect is the expansion and collapse of QV-bubbles within the surfaces. The particle aspect is created at interrelated points of encounter between the QV-bubbles in the expanding surfaces. The quantum vacuum in a QV-bubble contains vacuum energy that is injected into the bubble by the thrusts of oscillating strings.

A good illustration of how oscillating thrusts can create a continuous reality is the way music is produced by using a *bagpipe*, an instrument that is popular in Scotland. The continuous sound of the music is energized by air forced from a leather bag by pumping motions of the arm. The leather bag is kept filled with air by the breath of the musician playing the bagpipe. Here we have separate thrusts of energy that are fed into an uninterrupted process: the continuous sound of the music.

The air in the leather bag of a bagpipe is like *potential energy*. It comes into play, in the true meaning of the word, when the musician releases the compressed air through the appropriate openings in the pipes of the instrument, where the vibrations of the musical sounds are created. Potential energy is energy that remains inactive and unobservable until it causes an event. Here, the event is the creation of a musical sound.

The concept of potential energy is important in the context of the events examined in this book. At the top of the list, these are the *anomalous events* that cannot be explained by the known laws of physics. Also on the list are the processes that cause "normal" events but are poorly understood. As physicists readily admit, one unsolved mystery is the phenomenon of gravitational attraction. Another one is the process that causes the opposite effect: the force of repulsion that makes the universe expand. This includes the explosive event called the big bang. That event created the physical reality in which we exist. It also created our consciousness.

The force of expansion in our universe is like "cosmic anti-gravitation", as the journal *Science* put it, according to the German newsmagazine *Der Spiegel*. Astrophysicists have suggested that this energy may be another one of the universal constants, a "cosmological constant". The German science journalist Stefan Klein explained in *Der Spiegel* that this energy exists in a vacuum, in a physical "nothingness" that "pushes against its boundary, as if seeking to enlarge the space in which it exists".(10)

This supports the scenario proposed here. The energy of expansion is a universal constant relative to the observer of that expansion. The observer's qualia of consciousness exist in the expanding vacuum of a QV-bubble. The "nothingness" corresponds to what Manthey (3/3) called the initial void, and what Zizek (3/4) described as the hole in physical reality that is filled by phenomenal experience, i.e. by the elementary ingredients of consciousness. These ingredients are the qualia that are created by oscillating strings.

The Physics of Encounter suggests that the big bang was an encounter between two virtual realities. What the astrophysicist Hasinger (4/6) called the "spark that ignited the big bang" was a *collision* of gigantic potential energies. They were the vacuum energy in VE-surfaces expanding toward each other. In principle, it was the same type of collision that physicists produce in today's particle accelerators. The collision that occurs in a particle accelerator is like an explosion that results in countless newly created particles. Highly compacted energy is scattered into all directions of space.

The spark that ignited the big bang, as Hasinger said, could be the mysterious "dark energy" in our universe, "a force of repulsion that pushes the galaxies apart". If we can identify the source of this hidden form of energy, he argued, this "may shake the foundations of physics". The foundations of physics are the so-called *standard model* that describes the particle structure of matter and the forces that act upon the particles or originate within them. The physicists at the European Center for Nuclear Research, or CERN, near Geneva, spent 14 years and billions of dollars building a particle collider to put the standard model to its ultimate test. The goal, as reported by Dennis Overbye in the *International Herald Tribune*, was "to recreate energies and conditions last seen a trillionth of a second after the Big Bang" and to confirm the existence of a particle that is the linchpin of the standard model.**(11)**

"The standard model is what you could call a damn good theory", explained the science journalist Bas Kast of the German newspaper *Der Tagesspiegel*, "but one important piece of the puzzle·is missing. The theory fails to explain one thing, and unfortunately, that is a rather fundamental question. It is the question how matter acquires its mass. (...) How come there are particles that are not very heavy, like electrons, whereas other particles, like protons, are much heavier?"**(12)**

The effect of the missing particle, according to the standard model, is the answer to this question. If the standard model is correct, the particle has to exist. All of the previous experiments with particle accelerators, however, failed to establish its existence. The first start-up of the CERN experiment took place in 2008 but had to be discontinued for technical reasons. Irrespective of whether the experiment has taken place when this book is printed, the model of mind/matter interaction proposed here provides a description of the "missing particle" that is in keeping with the basic assumptions about its pivotal role in elementary physical processes. The Nobel Prize physicist Leon Lederman called it the *God Particle*, because in the standard model it is the key to understanding fundamental physical reality.**(13)** In the scenario proposed by the Physics of Encounter, however, this unique particle cannot be "localized" in the traditional sense because it exists in the hidden reality of quantum vacuum fluctuations. The details are explained in Part Two of this book, together with details about the role of the hidden reality in creating anomalous events.

The plans to simulate the big bang in the particle accelerator at the CERN facilities in Switzerland raised a question that seemed to come straight from a science fiction plot: would the experiment, because of the enormous energies generated in the accelerator, create *black holes* like the ones that exist in the outer regions of the universe? Black holes are points of such powerful gravitational attraction that all matter near the hole is pulled into it and disappears. Could this happen here on earth?

As reported by Dennis Overbye, two critics of the experiment filed a lawsuit requesting a restraining order to prohibit CERN from proceeding with the experiment until it produced a safety report and an environmental impact statement as required under the U.S. National Environmental Policy Act. A physicist who said he was a member of CERN's Safety Assessment Group confirmed that the matter was being reviewed. "The possibility that a black hole eats up the earth", he said, "is too serious a threat to leave it as a matter of argument among crackpots." According to some variants of string theory, black holes could indeed be created in a powerful particle accelerator. But the cosmologist Stephen Hawking had argued in 1974 that they would rapidly evaporate, and the possibility that they would appear at the site of the CERN experiment was widely ballyhooed in many scientific papers.(14)

Nonetheless, the lawsuit contended that miniature black holes might be created and then fuse, or grow, gradually developing the capability to swallow Earth. If the experiment has taken place when you read this, we can probably breathe a sigh of relief. We need not worry that destructive black holes can be created by ambitious physicists. With this tongue-in-cheek comment I do not want to disparage the theories of black holes. The theories support the view that our universe contains two different kinds of physical reality. One is the reality we observe and that physicists can measure. The other reality is hidden energy. The two realities originate at the two endpoints of oscillating strings and are inseparably intertwined. Scientists investigating anomalous events have called the intertwined realities *entanglements*.

*NOTES*

(1) *International Herald Tribune* (2008), p. 6
(2) VanWijk (2001), pp. 188 and 193
(3) Leonhard (1974), p. 11
(4) *Der Spiegel* (1986), p. 183
(5) Klein (1998), p. 179
(6) Folger (2003), pp. 39 and 41
(7) Arp (2000), p. 448
(8) Hasinger (2003), p.214
(9) Manthey (1997), pp. 2-4
(10) Klein (1998), p. 179
(11) Overbye (2008), p. 2
(12) Kast (2008), p. 32
(13) Lederman (1993)
(14) Overbye (2008), p. 2

# *Chapter 5*

**Hidden energy and morphic resonance. The effect of staring at someone. The smile of the Cheshire cat and the "extended mind". Converging hemispheres and the big bang.**

How much hidden energy spills over into the world of observable events when entanglements between the two realities in our universe create anomalous events? The Physics of Encounter suggests that the hidden energy makes physical objects move. It is the force of gravity, of which Einstein said that it is not a "force" as generally understood, but the "geometry of space". As I explained, gravity effects are generated at points of encounters between the expanding surfaces of VE-spheres. The points of encounter are miniature black holes that appear and disappear instantly as they energize the interactions between elementary particles. Things are different, however, during anomalous events.

The worry of some physicists regarding the ambitious experiment at the CERN laboratory in Switzerland was that the enormous energy released by the high-speed collisions between elementary particles might create black holes that would then grow and become dangerous. If one of the black holes were to become as massive as some of those that exist at distant locations in the universe, it was argued, the gravity effect could pull the entire planet earth into the hole. Our planet would cease to exist in the "here and now".

Where would it be? The answer: elsewhere in time, relative to the universe we would be observing if we had remained in the time frame that defines our present moment. What happens when a black hole "grows" and becomes more massive? The hole is an invisible location in space. Astrophysicists know that black holes are "out there" because of their abnormally large gravity effect, which influences the motion of the stars and planets in their vicinity.

A black hole that has grown can be visualized as a point of encounter that is superposed on many other points of encounter, as described in the preceding chapter. If the dimensionless points were visible, we would nonetheless see only one of them, because all the others would be "behind" the one point we see. The black hole is like a line of points that extends directly away from us. We cannot see the line, only (hypothetically) one of its endpoints. All points in the line are points of encounter between QV-bubbles in the surfaces of expanding VE-spheres. All points, therefore, are points where gravity effects are created. The more points there are, the stronger the gravity effect.

The points are aligned because they are the endpoints of thrusts occurring in resonance. The thrusts are the oscillations of strings in the VE-surfaces encountering each other. The synchronized thrusts are like the combined energy in a taut rope clutched by participants in a "tug of war". The synchronized and therefore highly amplified thrusts create the waves in the interacting VE-surfaces, and the waves "enmesh" because they are in resonance. The equal and opposite thrusts in the interacting VE-surfaces cancel one another in all three dimensions of space and the surfaces collapse. A new VE-sphere is created, at the point of encounter, by the combined effects of the thrusts.

In the above scenario, a gravitationally powerful black hole in the universe is an abnormally long line of superposed points of encounter that corresponds to the dimension of time. This scenario is in keeping with the arguments of various physicists that other universes exist inside black holes at other points in time. The string theorist Gabriele Veneziano, for example, argued that inside a black hole, "space and time switch roles. The center of a black hole is not a point in space, but a point in time" where a big bang may create a new universe.(1)

According to Lee Smolin, our universe continuously gives birth to many additional universes through black holes.(2) Veneziano reasoned that our own universe was created in the center of a black hole. "While the incoming physical matter was approaching the center, the matter density increased. As soon as the density (...) reached its maximum value, the process was reversed (...). The moment of reversal was what we call the big bang. The inside of one of these black holes became our universe."(3)

Veneziano's reasoning supports my argument that the big bang was an encounter between accumulated virtual effects in the surfaces of VE-spheres expanding toward each other in opposite directions of time. The virtual effects in VE-surfaces become measurable physical reality at interrelated points of encounter that include the location of an observer. At any one instant, the observer's location is the point of encounter where the qualia of the observer's consciousness are created through the effects of oscillating strings.

Such encounters can be visualized as encounters between the *hemispheres* of two VE-spheres expanding toward each other. The other two hemispheres of the two interacting VE-spheres expand away from each other. In the scenario proposed by the Physics of Encounter, hemispheres expanding *toward* the observer create what the observer experiences as a *matter effect*. Hemispheres expanding *away* from the observer correspond to effect of *anti-matter*. Whether the effects of VE-surfaces are matter effects or anti-matter effects depends on the location of the observer relative to the particles created by these surfaces.

In a cosmic context, the effects depend on the location of the above-mentioned "planetful of people" (4/1) who were told that a black hole created in a nuclear research lab might yank their entire planet out of the currently existing universe. The normal state of affairs is that such powerful gravity effects are only created by processes of cosmic dimensions in space and time, in the outer regions of the universe.

The metaphorical image of a hemisphere contributing to the creation of the big bang, as suggested here, was also used by the cosmologist Stephen Hawking. Like other scientists, he argued that the big bang was triggered by a pre-existing reality. According to the science journalist Rüdiger Vaas, Hawking argued that this reality had the shape of a hemisphere.(4)

A hemisphere expanding toward the observer is not something graspable or measurable moving through space. It is like the smile of the Cheshire cat that puzzled "Alice in Wonderland". It contains the characteristics of something as a possibility, but the physical reality that acquires these characteristics is not (yet) there. Eventually, other hemispheres will contribute to the creation of such a reality at interrelated points of encounter. Before that occurs, the "smile" is there, but not the cat. I have already explained that the concept of probability waves used by theoretical physicists corresponds to the non-graspable reality of waves in an expanding VE-surface. The surface contains only the *possibility* of a physical event, but it is not the event itself.

*(The probability wave) meant a tendency for something.*
*It was a quantitative version of the old concept*
*of "potentia" in Aristotelian philosophy.*
*It introduced something standing in the middle*
*between the idea of an event and the actual event,*
*a strange kind of physical reality*
*just in the middle between possibility and reality.*
Werner Heisenberg

The physics of probability waves, or interacting VE-surfaces, illustrates one of the principles of dialectic materialism. The principle encapsulates the paradox that we can "get something from nothing", as Manthey put it, if the "nothing" is not-yet-something. Zero plus zero is not zero if two physically not yet existent realities are potentialities, or possibilities, that complement each other and "enmesh", thereby becoming an observable whole. The two complementary realities, in this scenario, are the hemispheres that inject their potentiality into the creation of a new VE-sphere.

One of the hemispheres contains the potential matter effect experienced by an observer, the other hemisphere is the anti-matter effect created by a conscious experience. One hemisphere expands toward the observer, the other expands away from the observer. Matter and mind are two sides of the same coin.

> *Are you listening to those stars? They must tell us something.*
> *Matter consists of waves. And the waves? Are questions.*
> *Existence is dialogue. Everyone is two - or nothing.*
> Ernesto Cardenal

This poetic image of our existence in the physical reality of the universe is both dialectical and religious. "Physical science and religion are no longer incompatible opposites", declared Johann Dorschner. He and other physicists with similar views point to the fact that even the slightest change in the laws of physics and the values of the universal constants would have prevented life from evolving anywhere in the universe. If gravity, the weakest of all physical forces, were just slightly stronger, neither suns nor planets would exist today. The universe would have collapsed shortly after the big bang. "It does seem as if we were part of the design from the very beginning", Dorschner said, according to the science journalist Olaf Stampf. "That might be the magic bridge between cosmologists and theologians."(5)

Stephen Hawking is one of the physicists with a different viewpoint. He applied the laws of quantum mechanics (particle interaction) to the universe as a whole. Hawking justified his scenario, according to Stampf's article, by arguing that when the universe began to expand, it was compressed into a space that was smaller than an atom. The article in *Der Spiegel* was dismissive of Hawking's arguments: "Lo and behold, from Hawking's mathematical equations a universe appeared like a rabbit out of hat, as a ball of energy from which everything else evolved - without any influence from the outside. It just appeared as a 'quantum vacuum fluctuation'. 'That does not prove that there is no God'," the article quoted Hawking, " (...)'it just means that God is not necessary'."**(6)**

The author conceded that the origin of the universe remains a cosmic puzzle. "The energy droplets created by the big bang became two opposite building blocks of matter, a particle and an anti-particle. The present laws of physics state that both types of matter were created in equal amounts and then annihilated each other in a fraction of a second. To this day, physicists are wondering why one type of matter survived the inferno - and formed stars, planets, plants, animals, and humans."**(7)**

The Physics of Encounter suggests that both types of matter exist in today's universe, in equal measure. Only one type, however, is directly observable, because it is the elementary physical reality that is *coming toward us*. It is the reality created by VE-surfaces expanding toward the points of encounter where our consciousness is created. The effect of the other type of reality, of anti-matter, is present all around us, as the *positive electric charge* of elementary particles. Positive charges are created at points of encounter with VE-surfaces expanding away from the observer.

The images of the processes creating elementary particles and electric charges, as proposed here, are supported by measurements showing that a positively charged proton, for example, has less mass than a neutron. With the exception of a proton's positive charge, the two particles are exactly alike. It's worth noting that, since the surfaces of VE-spheres collapse in their entirety when they are encountered at any one point, the creation of a negative electric charge involving one of the hemispheres can pre-empt the creation of a positive charge involving the other hemisphere, and vice versa. Details of the proposed classification and characteristics of particles based on the Physics of Encounter are presented in Part Three of this book.

The anti-matter in the universe manifests itself here on earth through its *effects* as positive electric charges, but not as the *mass* of veritable particles. Positrons, the positively charged counterparts of electrons, "travel backward in time", as the Nobel Prize physicist Richard Feynman observed. They instantly depart from the location where electrons are created and are, therefore, just a mathematical construct that balances the equations of what physicists call quantum electrodynamics. It was Feynman who came up with the specifics of that concept.

The search for answers about the processes that create observable reality is philosophically polarized. The dialectic materialism mentioned above, as a principle underlying the creation of elementary particles, is rejected by the school of philosophy called idealism. Adherents of this philosophy regard the role of the mind as the primary creative source. Many religiously inclined thinkers favor the philosophy of idealism because it provides a focus on the world of *ideas*, in particular the idea of a free will and of our moral accountability in accordance with the will of God. The image of the big bang is their instinctive preference because it implies an act of creation. Their sense of reality tells them that the mind trumps matter because the mind can recognize the limitations of matter. They point out that only the mind can produce total perfection. Images of geometrically perfect figures, for example, such as a perfectly round sphere, a perfectly level plane, or a perfectly straight line exist only as ideas. Physical objects provide nothing more than rough approximations.

The doctrine of materialism, on the other hand, states that *matter* is the primary reality and that all phenomena, including thought, will, and feeling, can be explained in terms of physical processes. Those who favor this doctrine argue that the totality of physical processes has no beginning and no end. This makes it unnecessary to assume that the universe owes its existence to an act of creation such as the big bang, or to attribute the creation to a supernatural force such as God.

Can matter events of appropriate complexity, occurring in a human brain, bring forth consciousness and feeling, like the joy of experiencing "bit bangs", the expansion of knowledge? If so, the doctrine of materialism could be right. Is the source that creates the things we observe in our universe something that is not made of matter, but exists as forms, patterns, and images of the kind that arise in our mind? If so, the doctrine of idealism could be right.

The Physics of Encounter puts mind and matter on an equal footing. Experiments that suggest the existence of probability waves confirm one of the principles of dialectic materialism, but also make it very clear that we must abandon the notion that the mind plays only a subordinate role in elementary physical processes. The experiments involving elementary particles provide ample evidence that acts of observation contribute to the creation of what is observed. This, too, is a perfect illustration of a dialectical process.

An act of observation, in this context, does not mean that a physicist actually looks at a particle. The concepts in the theory of quantum mechanics, as explained in chapter two, are metaphors, images that help us visualize processes that cannot be seen. A particle is a quantum of energy, a measurable amount of energy. In quantum mechanics, an act of observation is a *measurement*. What the science journalist Olaf Stampf called an "energy droplet" (5/7) is not something that the human eye can see.

Measurable energy is not involved in the anomalous events that can be triggered by actually looking at something or someone. The process, as interpreted here, involves the expanding surfaces of VE-spheres. Prior to an encounter, the expansion is a *virtual* event that occurs without the elapse of time. An *actual* event does not occur until the expanding surface is encountered. When the mind is active, events occur not only in the brain. As the physicist Wolfgang Pauli said (3/6): "From an inner center, the psyche seems to move outward (...) into the physical world."

The "psyche", or mind, expanding into the physical world as an invisible, non-measurable energy, can influence matter. That includes the elementary particles in another person's brain. In that way, the events in your mind can influence the events in another person's mind. You may have noticed that when you stared intently at someone who was not aware of your presence, that person somehow felt you were doing so. The person would then turn around and respond with a questioning look: why are you staring at me?

The explanation proposed in the scientific literature is that our mind exists not only in our brain, but also in the space that surrounds the brain. We have what researchers of anomalous phenomena call an *extended mind*. It does not just *receive* information about events in space. While this happens, it also *sends* information into space. When we look at persons or things, our mind acts like an extended field of influence that sends information into space.

Information is transmitted across space in the form of wave patterns. The wave patterns of light waves contain information. The light waves allow us to see the characteristics of physical objects. In the case of anomalous phenomena, the wave patterns are not measurable energy. They are the layers (superpositions) of expanding and collapsing QV-bubbles in the surfaces of VE-spheres interacting at points of encounter. These wave patterns are the quantum vacuum fluctuations of space.

Vacuum energy is not observable energy. It is potential energy. It contains the possibility of creating observable events. When the wave patterns of interacting VE-surfaces are in resonance in a configuration called a coupled map lattice, potential energy is transformed into measurable energy. That is the energy of particles interacting in the atoms of physical objects, including the atoms of the brain.

Experiments have confirmed that staring at someone can affect the other person's mind. With rigorous safeguards against any unintended sensory cues, various scientists, foremost among them the biologist Rupert Sheldrake, have conducted thousands of trials with participants who were asked to "guess" when they were being looked at by another participant who was kept out of view. The participants who were being looked at guessed right much more often than would be expected by pure chance. Sheldrake pointed out that the minds of animals are also receptive to that kind of effect. "Many pet owners have told me that they can attract the attentionof their animal by looking at it. If it is asleep they can wake it up by their gaze."(8)

What "information" is contained in the effect transmitted by the mind? For persons who become aware that someone is staring at them, it is not information in the usual sense. It is just a vague feeling of unease that there is some outside influence impacting on their mind. Is there some sense of the direction from where the influence comes? Sheldrake says yes. The following examples of how people report their experiences are from one of his surveys.

"Recently I felt someone looking at me from behind about 50 feet away and I turned and looked directly at the person without scanning." - "I was attending a lecture when 15 minutes into the program I felt a prickle and uncomfortable. When I turned around, seven rows back I found my husband's ex-wife staring at me." - "In a crowded market in India amid hundreds of people I felt a pull to turn around and saw an old woman staring at me. I felt I knew her and she knew me but she was a stranger."(9)

Why would someone experience the eye gaze of another person when that person is a complete stranger? What "resonance" was there between the old woman and the person who felt the effect of her stare? Resonance is necessary. Otherwise the effect does not occur. Let's turn first to Sheldrake's experiments. The participants who were stared at and those doing the staring were chosen at random and did not necessarily know one another before the experiment. There was no particular "bond" between them.

The answer to the puzzle lies in the fact that *all* events in the world of quantum mechanics, of interacting particles, are unpredictable, random events. As interpreted here, all of these "normal" events are rooted in an elementary resonance phenomenon. The events are an emergent, microscopic reality that arises from the resonance between the wave patterns of quantum vacuum fluctuations. The particles that unpredictably disappear and re-appear, or make quantum jumps inside atoms, are produced by encounters between the expanding surfaces of VE-spheres. The encounters occur if there is a resonance between the wave patterns of the surfaces. The wave patterns are QV-bubbles expanding and collapsing in the VE-surfaces while the VE-spheres expand. The QV-bubbles contain the elementary ingredients (qualia) of consciousness.

To let an "abnormal" (*anomalous*) event occur, therefore, all you have to do is to "let things happen" on their own. Such an event would have to be like an absent-minded glance at whatever has unexpectedly appeared, visually, in your mind. You just see what you happen to see. You do not think about anything else. As soon as you start to think about what's happening or allow other thoughts distract you, you reduce the probability that the anomalous effect will occur. You reduce the probability of an encounter between VE-spheres because you add additional qualia to your moment of consciousness. This changes the relatively simple wave pattern that exists when events in the world of elementary particles (*quantum mechanical events*) occur "on their own". The additional qualia are contained in QV-bubbles that are added to the surface of a VE-sphere. Now the wave patterns are more complex and less likely to resonate at points of encounter.

An example of quantum mechanical events that occur on their own is the exchange of photons between atoms. The emission and absorption of photons is a light wave. There is "resonance" in a light wave because the energy of a photon absorbed into an atom is precisely the same amount, or *quantum*, of energy that is emitted from a different atom. A photon is absorbed into an atom only if there is a precise "fit" into the energy pattern of the waves (electron "orbits") that surround the nucleus of the atom.

Photons - light waves - allow us to see what happens in the world around us. Their energy impinges on our eyes and causes the events in the atoms of our brain. Light waves are different from the waves that exist in the expanding surfaces of VE-spheres. But the principle is exactly the same. The *extended mind* can influence the atoms of another person's brain simply by *seeing* that person, provided there is a "fit", or a resonance, that corresponds to the absorption of a photon. This requires a specific state of mind, as described above.

Mind events do not influence another person's brain through photons. They do so through the vacuum energy contained in the expanding surfaces of VE-spheres. The expansion occurs instantly, without the elapse of time. As Briggs said (1/2): "Time does not enter the picture until a quantum system interacts with its surroundings."

When you come right down to it, the phenomenon is not really "anomalous". It does not violate the laws of physics, as interpreted here. It is in keeping with the unpredictability of quantum mechanical interactions. As with many of the phenomena that are deemed anomalous, scientists can prove that they exist only by resorting to statistical methods, analyzing many trials in experiments repeated many times with many participants. These show that the occurrence of such phenomena is often highly improbable and that a single event is not reproducible at will. Small wonder, when you consider that what is required is a resonance involving a specific state of mind that is difficult to achieve through a conscious effort. Such an effort runs counter to the requisite state of mind described above, namely the *exclusion* of distractions, thoughts, and feelings that interfere with the purpose of the experiment. Anomalous events must be *allowed to happen* with a minimum amount of extraneous mental input.

That kind of mental state is most likely to occur in moments when your mind is at leisure, when you allow things to come into view by pure chance, wherever you are, whatever is happening around you. We can assume that this was the case with the people who were mentioned in Sheldrake's surveys. There was a brief bond between the old woman who stared and the person who became aware of her stare: "I felt I knew her and she knew me but she was a stranger". The temporary bond was a natural quantum mechanical bond, so to speak, mediated by the quantum vacuum fluctuations of interacting VE-spheres. The process corresponds to the bond between atoms linked in a light wave through the exchange of photons. In the Physics of Encounter, all particles, all quantum mechanical events, including those in the atoms of human brains, are interrelated points of encounter between the surfaces of VE-spheres.

The "bond" investigated by Sheldrake and others, as interpreted here, is established because the emission and the absorption of a photon is, paradoxically, a *two-way process*. Obviously, a photon has to be emitted from an atom if it is to be absorbed into a different atom. But the converse is also true. The absorption into an atom is necessary for the emission from a different atom! The absorption "causes" the emission, so to speak.

A so-called *offer wave*, or advanced wave, which results in the absorption of a photon into an atom creates a so-called *echo wave*, or retarded wave. The echo wave results in the emission of an identical photon from a different atom. This paradox is one of the many *dialectical aspects* of quantum mechanics. The underlying processes, as interpreted by the Physics of Encounter, will be described in the following chapters.

Photons, contrary to earlier theories, are not just waves of light traveling through open space. As explained in the preceding chapter, more recent research suggests that so-called biophotons are exchanged between the atoms of the brain. This means that the vacuum energy of VE-spheres impacts directly on the atoms of the brain. When the resonance between wave patterns allows the energy of photons to be absorbed in this way, the energy influences the mind events created in the brain. This is the basis for *extrasensory perception* (ESP). Mind events can be influenced by outside events without involvement of the sensory organs.

To describe the natural bond that allows these and other anomalous events to occur I will use the term *morphic resonance*, which was introduced by Rupert Sheldrake, the biologist who did extensive research on the effects of staring at someone. The word morphic is derived from the Greek word for shape or form. Morphic resonance, as interpreted here, exists when the wave patterns of VE-surfaces are identical, allowing VE-surfaces to interact at points of encounter.

Amazingly, just *imagining* that you are seeing someone can influence the events in the other person's mind. This was demonstrated in experiments done by Schwartz und Russek. Instead of directly staring at someone, the participants in these experiments took turns imagining, with closed eyes, that they were intently looking at a designated person. With a success rate significantly above what would be expected by random guessing, the persons sensing the imagined stares were even able to indicate whether the imagined stare was directed at their head or the small of their back.(10)

Morphic resonance produces particularly startling effects in the minds of people who have a close relationship. The actress Shirley MacLaine, a close friend of the actor Peter Sellers, experienced such an effect and described it as follows: "I was sitting with some friends in my apartment in Malibu. I had been traveling and didn't know that Peter had had another heart attack. We were chatting amiably when suddenly I jumped up from my chair. 'Peter', I said. 'Something has happened to Peter Sellers.' When I said it, I could feel his presence. It was as though he was right there in my living room watching me say it. I felt ridiculous. Of course, all conversation stopped. Then the telephone rang. I disguised my voice and said hello. It was a newspaper reporter. 'I'd like to speak to Miss MacLaine', he said. 'Well, actually I wanted to get her reaction.' 'To what?' I said. 'Oh', he said. 'If you haven't heard, I'm sorry, but her friend Peter Sellers just died'."(11)

Shirley MacLaine's experience is a striking example of what researchers of anomalous events must take seriously and try to understand. There are many reports of similar experiences. In World War II, wives of soldiers woke up in the middle of the night with the horrible feeling that their husbands had just been killed in combat. They later found out that this was indeed the time when their loved one had died. An experiment illustrating this type of morphic resonance was done by Russian researchers. They killed mice that had just been born and measured the physiological reaction of their mother. The experiment made use of the fact that distress changes the electrical conductance of the skin. To test whether an anomalous influence is transmitted instantly over large distances, the mother was put into a submarine at sea. At a specified time, the young mice were killed in the laboratory. At precisely that moment, the skin physiology of their mother distinctly showed the symptoms of distress.

What reasonable conclusion can we draw from such an experiment? Did the fledgling minds of the young mice instinctively send out something like an SOS signal that was picked up by their mother? Did the adult mouse have a mind that extended across the expanse of an ocean? Neither assumption seems entirely appropriate. We cannot be sure, for example, whether the freshly born little creatures possessed anything that could be defined as a mind. The Physics of Encounter suggests that the results of the experiment may have little or nothing to do with the effects of *mind* events, but rather with *matter* events that occur when a biological system dies.

This means that Shirley MacLaine's experience involving Peter Sellers did not necessarily result from telepathy. The actor may or may not have thought of her when he died. Perhaps he died in a coma. The event in Shirley MacLaine's mind, like the physiological reaction of the mouse whose offspring were killed, may have been triggered by a distant matter event: the transition from life to death. The processes that make matter *alive* are one of the issues examined in Part Two of this book.

The energy involved in brain events triggered by morphic resonance is minimal. But the energy released by other anomalous events can be enormous. Cases in point are the power output of the light emitted by UFOs and the sometimes extremely heavy objects that hurtle through the air in poltergeist events. The process releases some of the energy that is hidden in the pre-manifest reality of quantum vacuum fluctuations. Ervin Laszlo described the energy this way: "It is the originating source of matter itself. (...) The quantum vacuum contains a staggering density of energy. John Wheeler estimated its matter equivalent at 1094 gram per cubic centimeter, and that is more than all the matter in the universe put together. Compared with this energy density, the energy of the nucleus of the atom, the most energetic chunk of matter in the known universe, seems almost minuscule: it is 'merely' 1014 gram/cubic centimeter. (...) The 'real' world of matter, that is: of energy bound in mass, is (...) much less energetic than the vacuum (...).(12)

What, precisely, is mass? Part Two of this book will look into this question in the context of anomalous events. Before I get to that, I want to use the next chapter to show that science is not always a dispassionate quest for answers. There are bitter quarrels. Loosening up might help matters. The question whether the particles of light (photons), counter to previous assumptions, might possess a smidgen of mass, for example, was tackled with a sense of humor, albeit not from a member of the scientific establishment.

*Photons have mass?!? I didn't even know they were catholic.*
Woody Allen

*NOTES*

(1) Veneziano (2004), p. 38
(2) Smolin (2006)
(3) Veneziano (2004), p. 38
(4) Vaas (1998), p. 103
(5) Stampf (1998), pp. 166 and 169
(6) Ibid, p. 169
(7) Ibid, p. 169
(8) Sheldrake (2000), p. 231
(9) Ibid, p. 127
(10) Schwartz and Russek (1999), pp. 220 and 226
(11) MacLaine (1983), p. 173
(12) Laszlo (1996), p. 4

# Chapter 6

**Quarrels about the scientific method. Spoon-bending parties and Poltergeist events. Psychic research and the CIA. Did Lincoln attend spiritualistic séances?**

Intense intellectual battles are fought in the scientific community. I will pick one example to show that when it comes to the way the disagreements are phrased, the antagonists keep their scientific gloves on, but the message is clear nonetheless: the other guy has got it all wrong.

The philosopher Robert Almeder, in an essay on *reincarnation* published in a scientific journal, mentioned the case of a young American woman by the name of Ann Davis who became aware of what she experienced as strange memories. Based on these memories, she said, she must have been Napoleon in a previous life. A case for the loony bin? She is not deluded about who she is. She does not think she is Napoleon. But she is able to speak, in proper dialect, Napoleonic French, a language that she had never learned. She is not a very educated woman, but she correctly describes the battle of Waterloo, in great detail and in a way that only Napoleon's military mind could describe it. She tells of memories that only Napoleon could have had.(1)

There are numerous highly plausible instances (called "rich cases" in the scientific literature) in which a wealth of details clearly indicate that some people possess skills and knowledge they did not acquire through the normal process of learning, personal contacts, or practice. The daily routine and educational background of these people offered no opportunity for a fluency in a foreign language, or an ability to play a musical instrument with remarkable perfection. The explanation often given is that these skills were acquired during a previous lifetime.

Should we accept the concept of reincarnation as an explanation for such cases? Almeder said yes, based on his view that consciousness can exist without a brain and that a "sufficient criterion for personal identity is continuity of memory". He argued that the "human personality (...) survives biological death as the repository of certain mental states and dispositions having to do with memory, intelligence, sense of humor and acquired cognitive skills (...)". Almeder admitted that "it is difficult to imagine" how a human being can survive "as a causal agent in the world without having a physical body", but points out that "a bodiless person (...) in all likelihood would have some properties it shares in common with matter as we ordinarily understand it", so that "no law of nature would be violated" when a "bodiless person" acts as a causal agent in reincarnation.**(2)**

Almeder, referring in his essay to a review of his book "Death and Personal Survival", quotes the reviewer James Wheatly, who wrote: "I do not find anything in this book to counter (the) argument that the idea of a bodiless person is logically incoherent (...)." Almeder mentioned other critics of his arguments who wrote that he made an "overstatement" in connection with "audacious" and "extravagent" claims, but then Almeder goes out on a limb himself when he insists, in his essay, that it "is irrational to reject the view that some people survive, in some measure, bodily death as we know it. It makes no sense to deny it."**(3)**

The honorable practice in scientific battles is to mention the viewpoints of others who choose to disagree, but the acrimony overshadows what should be joint efforts to shed some light on controversial issues. The dispute is often not about the substance of issues, but about the concepts used to interpret what is broadly accepted empirical evidence. The concept of "psi", for example, is used by some researchers to designate what they regard as anomalous powers of the mind. Almeder rejected that concept and argued that, since the existence of *psi* is "neither confirmable nor falsifiable by appeal to any factual evidence at all (...) it is difficult to see any explanation couched in terms of it as anything more than a merely logically possible explanation, no different in kind than offering explanations in terms of angels (...)."**(4)**

Does "psi" exist? That's like asking: do the particles called "quarks" exist? They are assumed to be unobservable, "non-localized" (!) particles inside other particles called hadrons. The theory of quarks provides an explanation for the observed events involving hadrons. I will get to that later. Just this much here: the Physics of Encounter suggests that quarks are encounters between expanding VE-surfaces that occur inside more complex particles. Are there quarks inside hadrons? All we have to go on are what hadrons do when we observe them.

This brings me to the school of psychology called "behaviorism". I had a tough choice to make when I had to decide what graduate school to choose for a position as research assistent in experimental psychology. My professors had tended more toward "Gestalt psychology" and made no bones about their view that the adherents of behaviorism were too narrow in their approach to events inside the mind. Is there consciousness inside the minds of animals? I knew that the behaviorist school regarded this question as irrelevant, or let's say unanswerable through scientific experiment. I chose the University of Wisconsin at Madison, renowned for its experiments with animals. Animals cannot tell you whether or not they experience consciousness, but that has nothing to do with their learning curve and other quantifiable aspects of how they use their brain.

What made the decision tough for me was that during my studies, I was taken aback by the vehemence with which behaviorists and Gestalt psychologists, in their publications, disparaged one another. Quantification and math, I knew, will not give you the entire picture, as expressed by the German word *Gestalt*. The same goes for what philosophers of science call reductionism. We miss the essence of an all-encompassing reality if we reduce our study of consciousness to measurements of neurological events in the brain.

Physicists are on the wrong track if they insist, with cutting condescension, that only precise quantification will endow theories and research results with the halo of true scientific achievement. Summarizing "a host of new studies" regarding the correlation between mathematics and intuitive insights, the science journalist Natalie Angier wrote in the *International Herald Tribune* that "math teachers might do well to emphasize the power of the ballpark figure, to focus less on arithmetic precision and more on general reckoning."(5) This is in keeping with Jahn's admonition (3/8) that physicists should recognize the usefulness of metaphorical images in their efforts to enhance the understanding of the interrelationship between physical reality and consciousness.

The materialistic approach of many scientists bent on proving their point through quantification puts them on a collision course with researchers whose subject matter, the study of paranormal phenomena, often prevents them from coming up with comparable "proof". To the readers of this book who, I presume, are interested in finding out more about anomalous events, I will pass on this word of caution published in a book by Rosemarie Pilkington, an educator with a Ph.D. degree in consciousness studies: "Searching the internet can be very productive, but you may find avid, uncritical proponents of spiritualism as well as many professional debunkers (...). These latter groups, arch-skeptics, are as fanatical as some of the anti-psi religious fundamentalists who consider any psychic manifestation as the work of the devil."(6)

With charlatans and frauds eager to cash in on the need to believe in other-worldly forces, and an abundance of sensationalist reports tailored to "sell" a point of view or to sell newspapers, journals and books, it is no surprise that a skeptical view prevails among many who are more thoughtful and educated. They are put off by religiously colored fantasies hyped in fake spiritualistic sessions allegedly putting people in touch with loved ones who have died. The craze brings forth so-called mediums who make a living by claiming an ability to forecast the future for anyone who comes to visit them. Equally dubious to a critical mind are reports of otherworldly spaceships carrying godlike creatures who possess the power to save our world, or transporting cosmic imperialists with fiendish plans to destroy our civilization.

I have already indicated that, in spite of the many outlandish aspects of what less than reputable sources describe as paranormal reality, the public and scientists alike should recognize that a small portion of such reports warrants closer investigation. What is needed is a yardstick for determining what reports are plausible. The public interest in this issue is strong, but since even open-minded scientists are not sure what they are dealing with, the regrettable upshot is that the subject matter draws many people into two fiercely antagonistic camps. On the one side are the naively avid believers, and on the other side the equally avid, knee-jerk debunkers who throw out the baby with the bathwater.

The scientists who have decided to take a closer look are caught in the middle. Their research suffers from lack of funding because most reputable institutions consider such work an unworthy enterprise. The institutions, and the individual scientists fear that their reputation will be damaged if their research becomes known. Some researchers publish their theories and findings under an assumed name.

Let us look at a few more examples showing that some of the baffling events should be taken seriously. I already mentioned RSPK (recurrent spontaneous psychokinesis), which is commonly known as the Poltergeist effect. Since it is a recurrent effect, it often continues over a longer period of time (days or weeks). Psychologists investigating this phenomenon agree that it is caused subconsciously by the effect of troubled minds, by stress, frustration, deeply felt needs, or feelings of hostility. In several cases, investigators identified the individuals who unknowingly caused the RSPK effects and discovered what was troubling them. When the problem was addressed openly, in a mix of psychotherapy and changes in the personal situation of the individual, the RSPK effects stopped.

The physicist William Roll witnessed many Poltergeist events because he traveled to the location where they occurred as soon as they surfaced in the news and in police reports. He stayed there for meticulously documented observations and research in cooperation with psychologists and police investigators. Since the subconsciously triggered psychokinetic effects often cause considerable damage to valuable objects, police officers are usually called in to ascertain whether the damage was caused by a willful and malicious act.

One case described by Roll involved a 14-year old girl, Tina, who had been the center of poltergeist disturbances that were filmed by a news crew. Before Roll and a colleague could begin their investigations, "the activity around the girl had died down. To bring the incidents back, a psychotherapist (...) who was counseling Tina suggested that hypnosis might evoke the bodily sensations that had been associated with the events and thereby (...evoke...) the events themselves (...). This led to a resumption of the occurrences. (...) To study the incidents under controlled conditions, we set up a table with (...) targets (...for the poltergeist effect). (...) The heaviest target to move was a 12 inch socket wrench. (...) In all, there were 21 movements of objects when Tina was under observation, of which eight came from the target table." Roll and the other experimenter were caught by surprise when the heavy wrench suddenly moved. They noticed nothing when it apparently flew past them and past Tina, who was standing nearby. But then they heard a loud noise from the hallway behind the girl. "The wrench had hit the door to a storage room. It had moved 18 feet."(7)

Another case of RSPK described by Roll occurred in what neighbors had begun to call "the house of flying objects". A police detective and Roll were present when one of the RSPK events occurred. The 12-year old son of the family was initially suspected of trickery or spiteful acts, but could not have caused the event in that way because Roll and the detective observed it directly. Roll and an RSPK investigator concluded "on the basis of psychological studies that the boy had unknowingly caused the events. Most of the objects that moved and broke falling to the floor were located in the rooms frequently used by the boy's father. The boy was found to have strong feelings of anger towards his father."(8)

In a warehouse for novelty items, an 18-year old shipping clerk named Julio was suspected of purposely causing the breakage of merchandise and the disruption of business. Roll and a second investigator set up several objects as targets for the RSPK. On several occasions, in their presence, objects moved, "seven of these when we had Julio in direct view. It seemed as if he was rewarding our attention." In his article about these events, Roll mentioned the conclusion of a parapsychologist "that objects affected by RSPK are 'substitute objects' that represent people associated with the objects", and also mentioned that a psychologist "who analyzed Julio's responses to a thematic apperception test (TAT) and Rorschach pictures (...) said that Julio experienced the owner as 'phoney and cheating' (...)."**(9)**

When eminent scientists like William Roll publish their findings in a reputable scientific journal, that is more than enough reason to accept the hard-to-believe phenomena as factual. Another criterion for taking reports seriously is the unblemished reputation of an international newsmagazine and of the people it interviews. Here is what *Time* magazine wrote in 1984 in an article about "psychic weapons", meaning: the powers of the mind to inflict damage through psychokinesis and to explore distant military sites far beyond the range of sensory perception through "remote viewing". The article appeared when the so-called cold war between the USA and Russia caused concern about the weapon systems developed in communist-ruled Russia. It was the dominant country in what was then the Soviet Union (U.S.S.R.). The parapsychologist Russell Targ, as reported in *Time*, did not "specify whether the Defense Department, the CIA or both funded his psychic research programs", but said that "there was a 'multimillion-dollar' project, part of which focussed on 'remote viewing' experiments. (...) On a visit to the U.S.S.R. in October, Targ found that the Soviets had replicated some of the experiments he and his colleagues had reported in scientific journals. Says Targ: 'In the Soviet Union, psychic research is taken seriously at the highest levels.' Sighting submarines by clairvoyance? (...) Representative Charlie Rose, a North Carolina Democrat on the House Select Committee on Intelligence, says it may be worth a look. 'Some people think this is the work of the devil,' says he. 'Others think it may be the holy spirit. If the Soviets, as is evident, feel it is worthwhile, I am willing to spend a few bucks'."**(10)**

In 1996, the physicist Hal Puthoff confirmed the secret project in a scientific article entitled "CIA-initiated Remote Viewing Programs at Stanford Research Institute".**(11)** A large part of the documents had been declassified the year before. After circulating reports about some preliminary experiments he had started in 1972, Puthoff was approached by the CIA to set up a program to widen the investigations, which he conducted with Targ and others. At a scientific conference I attended in 2003, Puthoff described the successful experiments. Clearly, they are more than just mind games for theoreticians. The CIA still keeps some its documents under wraps. Puthoff revealed that "73,500 pages have been released in full, 17,700 have been denied in full, and 20,800 pages are in review".(12)

To come to another type of anomalous force, I will mention the name of the Russian president Boris Yeltsin. This force is somewhat less controversial and is routinely described as a psychic ability of certain individuals to cure medical problems where the traditional skills of doctors fail. In 1994, as reported by the German newsmagazine *Der Spiegel*, Yeltsin personally pinned a medal of honor on the dress of the nurse Yevgeniya Davitashvili. For years, she had secretly treated the top officials of communist Russia, among them party boss Brezhnev and foreign minister Gromyko. She was immensely popular in Russia because she had also demonstrated her ability in public. By just touching the patients with her hands, she reportedly made their tumors disappear in a matter of days.(13) While the communists were in power, her activities were officially condemned as boastful and "unscientific". They ran counter to the reigning dogma of materialism. In that philosophic doctrine, as I explained, matter is the primary reality, not the mind. Yeltsin, as the first democratically elected leader of Russia after the collapse of communism, made a point by officially recognizing a force of nature that had previously been a taboo.

Demonstrations and claims of similar abilities by various "healers" have made headlines in many countries. A number of recent experiments by Moga and Bengston with laboratory mice have confirmed that such anomalous healing effects are possible. The mice were injected with tumor cells and some of them were then repeatedly given hands-on treatments by laboratory assistants. All of the mice that were not treated predictably died. Many of those that had received the unconventional treatment survived. Their tumors disappeared completely.(14)

One last example I will give here is the bending of a fork or a spoon through the power of the mind (psychokinesis). Demonstrations of this ability made Uri Geller famous. Apparently, it is not a hoax. According to Rosemarie Pilkington, it was done many times at "metal-bending parties". To produce the phenomenon, she explained, "it is of primary importance to prevent the conscious mind from interfering (...)", for example "by occupying it with 'noise and nonsense' (...)." Pilkington mentioned "the many 'spoon-bending' parties in the 1970s and 80s", and wrote: "I took part in one such party in Cambridge in 1982 (...). After a banquet dinner with wine, a group of us (...) met in a large comfortable room with the goal of bending cutlery. (...) As we sat around relaxed and in high spirits, the leader told us what we could expect. We were supposed to chant, 'Bend, bend, bend!' while lightly massaging our forks and spoons. (...) With his words firmly implanted in our minds, we laughed and joked and chanted and rubbed (...) and suddenly I got the feeling that I could bend my spoon. I tried it and, lo and behold, it 'gave' as though it was a licorice stick, then immediately hardened again. I looked at it. Sure enough, there was a slight bend in the handle. I held it up and shouted exitedly, 'Look, it bent! I bent it!' The floodgates broke. Everybody's flatware started to bend."(15)

Pilkington mentioned the researcher Dennis Stillings, who conducted a metal-bending party with a group of medical professionals. "It was a small group of primarily skeptical people who were into deep relaxation technique... After they shouted, 'Bend!' three times they went into a concentrative meditative state. After ten minutes passed with nothing happening, Stillings, in his own state of embarrassment, decided to do a little 'artifact introduction' in order to 'break state'. He saw that the fork being held by the person in front of him looked as though it was a little bent. (Perhaps he had not straightened it out very well before the party.) He exclaimed, 'Look! We're getting some bending!' That broke the unproductive meditative state and people started talking to each other. Soon forks and spoons were bending away. One physician stroked the side of a knife blade and became excited as he saw it begin to droop."(16)

A psychiatrist, in Pilkington's account of that session, took a heavy fondue fork with an ebonite handle and began to rub it. "Then suddenly the fork exploded. Fragments of ebonite struck against the windows and walls. (...) What had happened was that the part of the metal shaft imbedded in the ebonite had bent, even though it hadn't been touched directly, shattering the brittle ebonite. The now exposed shaft was bent about 20 degrees. Some time later Stillings took it out to show to someone and noticed that the bend was at a much greater angle than it had been at the night of the party. The metal had continued to bend in storage."(17).

Reports like these leave room scientific quibbling because so much seems to depend on the appropriate mood, the setting, and the individual "talent".

Similar reservations are voiced with respect to the UFO phenomenon. If the calculations of reputable physicists can be believed (and why shouldn't they?), the extraordinarily high optical power output of UFOs is indeed inexplicable by the currently known laws of physics. But UFOs cannot be summoned at will, and the customary scientific verification by any interested scientist is not possible.

This brings us to the highly controversial question about what constitutes "proper" scientific methodology if the goal of research is enlarged to include paranormal phenomena. The core of the problem is that, in principle, these seemingly inexplicable phenomena are like quantum mechanical events, which cause particles to appear and to disappear unpredictably. The quantum mechanical conundrum no longer bothers scientists, although Einstein was initially upset and exclaimed: "God doesn't play dice!" Scientists can work with the theory of quantum mechanics because the sum total of countless unpredictable events in the *microscopic* world of interacting particles creates predictable effects in the *macroscopic* world of observable objects.

That is not the case, however, when the hidden energy that causes "normal" particle events is suddenly and steeply increased through the unpredictable morphic resonance between the wave patterns of interacting VE-surfaces. These wave patterns, as explained in the preceding chapter, are probability waves. Highly improbable macroscopic events, therefore, can suddenly and unpredictably occur if the effects of human minds interact in morphic resonance and jointly unleash anomalous energy effects.

The search for an expanded scientific approach that takes into account the need to study anomalous events must, paradoxically, orient itself along the narrow approach taken by the behaviorist school of experimental psychology. As I was told when I entered graduate school, what counts is what animals and humans do with their brain. Therefore, acceptable scientific data about what humans do, how they behave under certain circumstances, can include what humans *say*. Among the things they can be asked to do is to describe what they felt or experienced. The statements can be duly noted and compared with what other people said under similar, if not laboratory-controlled circumstances. That is scientific evidence. Paradoxically, this seemingly restricted approach is well suited for enlarging the scope of investigation into anomalous, mind-related phenomena.

If a woman by the name of Ann Davis says she experienced what seemed like genuine memories persuading her that she had been Napoleon in a previous life, let's look at what other people have said about similar experiences and compare notes. Let's look at their behavior, their inexplicable skills. "Proving" that this is reincarnation is another matter. Other interpretations of the evidence are possible.

What all this boils down to is to take a second look at the distinction between private knowledge and public knowledge. Almeder, in his discussion of the anomalous mind force called psi, said that we can have "private knowledge of the existence of psi (...) based on evidence that is quite transitory, nonrepeatable and hence accessible only to the subject for a limited amount of time. But private knowledge is, by definition, not the public knowledge we seek in natural science, and there is no reason for anybody to accept an item of private knowledge as an item of public knowledge."(18)

I disgree. We must admit "private knowledge" into the investigation of anomalous events. Published and verified versions of what people have said, and the number of people who said the same thing, should be accepted as data for scientific theories (not, of course, as scientific proof). Experiments have shown that people feel they are being floated into the air against their will if a magnetic force is applied to their brain. The statements show that these sensations are similar to the ones described when people say that they experienced what seemed like an "abduction" by alien beings who arrived on earth in UFOs (unidentified flying objects).

In cases where several people report that they participated in a group session, or séance, during which all of them saw strange events involving real objects that seemed to be caused by some invisible force, investigators need to decide whether to accept such statements at face value, or to reject them, on principle, as unverifiable. There is no way of being absolutely sure that the statements are true. They could be "private knowledge", a group-generated delusion, for example, or they could be a plot to deceive the public. Even if the statements are true, they are like private knowledge in the sense that participants cannot provide a public repetition of the events upon request, for reasons described above.

In some cases, the alleged events were filmed or photographed, but that raises the question whether the cameraman or the photographer are reputable. They could be a part of the plot. For smoother sailing in difficult waters, scientists need to abandon the criterion of repeatability in certain cases and adopt a more flexible approach. They could achieve their goal dialectically, arriving at a high degree of *certainty* by reaching a consensus on the *probability* that the examined data are correct. It should not be difficult to agree on the reliability of statements that in all likelihood were made in good faith and with some knowledge of the pitfalls of mental processes. The criterion to be applied is the reputation of the people making the statements (education, scientific background, public standing). This approach is a justifiable reliance on what I will call *reputable sources*. It includes the fact that seasoned pilots are among the many witnesses who have said that they saw UFOs and observed their anomalous effects. (It's worth mentioning that president Reagan and president Carter also reported seeing a UFO.)

A second criterion is a consensus on what should be considered a *reasonable conclusion*, for example a conclusion drawn from public events that have a bearing on the subject matter to be investigated. The two criteria are interrelated. Many people in different countries who said they saw UFOs in the sky at night also mentioned that mysterious circular patterns ("crop circles") appeared overnight in the nearby fields of farmers. Are enough of them credible witnesses? Is it a reasonable conclusion that the crop circles can't all be hoaxes, since they reportedly appeared noiselessly, overnight, in many different countries? Could there be that many hoaxters all over the world? Could they deftly accomplish what engineering students in a published experiment were unable to do: the noiseless creation of such patterns, in the dark and in a matter of hours? What about the strange behavior of dogs, reportedly whimpering in apparent fear during a UFO phenomenon, vomiting when taken into a crop circle?

Here's another example of a reasonable conclusion. For years, the US government vigorously denied having any documents relating to UFOs. It later had to admit that this was untrue. After law suits based on the Freedom of Information Act, some of the documents were released. At the very least, this shows that the reports about UFOs were a matter of public concern, and not dismissed as isolated mental quirks or figments of the imagination. (The government of Canada, by the way, in a study published in 1954, concluded that some of the sightings involved "real objects".)

At this point, I will address the widespread and fully justified skepticism about the gatherings called séances, during which like-minded people sit around a table in a darkened room to conjure up disembodied spirits or to make the table rise from the floor. The reports about what happens during séances are indeed hard to believe. A criterion that is applicable here is a consensus on what are reputable sources. A good example is an article in the *Cleveland Plain Dealer*. According to an account in *Phänomene*, a reference work on anomalous phenomena, the article appeared while Abraham Lincoln was president. In it, Lincoln described some of the experiences he had when he participated in séances. When asked whether the article was true, he reportedly replied: "It mentions only half of all the amazing things that happened."(19)

It is possible, of course, that the reports about what Lincoln said are wrong. True, the article appeared in a reputable newspaper, not a tabloid. But reports need to be taken with a grain of salt if they are just hearsay, like Lincoln's alleged remark about the article. It is also possible that charlatans decided to dupe the president of the United States. Should we consider such chutzpa a likely possibility? We must let the facts speak for themselves. The newspaper article is a fact. It's understood that reasonable people may disagree on what are reasonable conclusions. But a few additional examples of reasoning by reputable observers and researchers may tip the scale in favor of taking such reports seriously.

Many persons of high public stature have participated in séances, among them, in the 1850's, the king of Prussia. According to Pilkington, the French writer Victor Hugo was also an enthusiastic sitter, and the American Poet William Cullen Bryant was among the many people from all walks of life who made attested statements about their spectacular experiences as witnesses of the anomalous force unleashed in séances. Among the séance participants mentioned in Pilkington's book are also university professors, a High Court judge, a psychiatrist, a playwright, and a journalist who later became prime minister of Iceland.(20)

None of this is sufficient evidence to dispel lingering doubts, but one additional factor should be taken into consideration: all of what is reported by a broad spectrum of serious-minded people willing to attest what seems preposterous fits into an explanatory pattern that is not at all preposterous. The proposed scenario that describes the processes underlying the reported events is based on theories that clarify fundamental questions regarding the nature of physical reality. These are the theories of oscillating strings and of quantum mechanics, integrated into the overarching explanatory pattern I have called the Physics of Encounter.

I have avoided the word "theory" in my description of the Physics of Encounter. It is a model, a scenario using metaphorical images. In one respect it is like a theory. It can be tested by experiments. I have proposed some at the end of this book. Theories, according to conventional wisdom, have to be based on empirical evidence, which is evidence based on experiments and observation, rather than theory. The philosopher Robert Almeder staunchly defends the time-honored tradition of scientific methods. A theory, he insists, is only acceptable if it is developed from empirical evidence and is *falsifiable* (can be proven wrong by experiment). In view of this, it is all the more remarkable that he uses the concept of a "bodiless person". With that provocative concept, Almeder lends support to the model of mind/matter interaction proposed here. There are, I am happy to say, many physicists, psychologists, and philosophers whose thinking runs along similar lines.

Almeder's critics, like Stephen Braude, argue that falsifiability is not necessary for a reasoned decision to work with a theory until someone comes up with a better one. The first hurdle to be taken in developing a theory is to come up with an *hypothesis*, an unproved theory temporarily accepted to provide a basis for further investigation. An hypothesis is the initial effort to explain a *phenomenon*, which the dictionary defines as "any event, circumstance, or experience that is apparent to the senses and that can be scientifically described or appraised". An anomalous event is a recognizable phenomenon because it has been scientifically appraised as not in conformity with the known laws of nature. But it is, for that very reason, a decidedly *intractable* phenomenon, meaning: it is very difficult to "get a handle" on it. No wonder, since anomalous phenomena seem to be caused by processes that are not an observable reality. The processes exist in a world where (counter to our conventional logic) an event can occur through retrocausation. The origins of time and space, in this "unreal" world that we cannot grasp or measure, exist as quantum vacuum fluctuations.

Stephen Braude, therefore, has a point when he writes (as quoted by Robert Almeder) that "the non-falsifiability of an hypothesis may simply reflect the intractable nature of the phenomenon in question, rather than a theoretical deficiency. (...) It would be (...) presumptuous to insist that nature operate only in ways amenable to the preferred methods of science."(21) A new approach, or better: an additional, more flexible approach, is needed. That approach must clarify under what conditions scientists should make use of what people say about their "private knowledge" (inner experiences) if it enriches the domain of public knowledge. This is particularly useful in the development of hypotheses that stimulate further research. It is already being done in many cases.

*We have found a strange footprint on the shores of the unknown.*
*We have devised profound theories, one after another,*
*to account for its origins. At last, we have succeeded*
*in reconstructing the creature that made the footprint.*
*And lo! It is our own.*
Arthur Eddington

*NOTES*
(1) Almeder (1996), p. 495
(2) Ibid, pp. 510, 511, and 515
(3) Ibid, pp. 507 and 516
(4) Ibid, p. 506
(5) Angier (2008), p. 10
(6) Pilkington (2006), p. 236
(7) Roll (2003), p. 78
(8) Ibid, p. 78
(9) Ibid, p. 76
(10) *Time* (1984), p. 9
(11) Puthoff (1996), pp. 63-76
(12) Puthoff (2003), p. 15 (Abstracts)
(13) *Der Spiegel* (1994), p. 235
(14) Moga and Bengston (2007), p. 21 (Abstracts)
(15) Pilkington (2006) pp. 206-208
(16) Ibid, pp. 208-209
(17) Ibid, pp. 209-210
(18) Almeder (1996), p. 505
(19) *Phänomene* (1994), p. 199
(20) Pilkington (2006) pp. 1-241
(21) Almeder (1996), p. 502

## Chapter 7

**Avid believers and fanatical debunkers. UFOs, spy satellites, and cover-ups. Prophetic dreams. The entanglement of the agent and the receiver. Clarifying relevant concepts.**

Many arguments of the so-called "debunkers", of scientists and other clear-minded thinkers who reject the unreasonable claims about paranormal phenomena, should be beyond dispute. But, alas, they are not. There are avid believers who are immune to rational arguments and scientific evidence. What is more surprising is that some of the debunkers, citing what in their view are scientifically established facts, are equally vehement and unreasonable in their rejection of the evidence that certain aspects of currently inexplicable events warrant scientific investigation.

UFOs, the "unidentified flying objects", and the alien beings who reportedly pilot them to earth, are a major part of the kaleidoscope of events that cannot be explained by the known laws of nature. Many of the effects that occur when UFOs are observed are also associated with the broad spectrum of other anomalous events. Examining the concepts used to identify such events can clear up some of the gun smoke in the battle about the pro and con of this issue.

Let me say at the outset that the wealth of undisputed evidence, interpreted in terms of the processes described in this book, points to clear-cut conclusions: (1) Yes, UFOs are physical phenomena. (2) No, they are not spacecraft piloted by alien beings. (3) UFOs are phenomena in the earth's atmosphere that can be picked up by radar and generate a variety of physical effects. (4) The phenomena are triggered by anomalous mind events that are powerfully amplified by the hidden energy in our universe. (5) The hidden energy kicks in when human minds interact in morphic resonance. (6) The energy is provided by the quantum vacuum fluctuations that pervade all of space.

In recent years, many researchers have begun to avoid the word UFO because they have recognized that what seem to objects in the sky are not the "nuts-and-bolts" reality of aircraft or spaceships. They now use the term *anomalous atmospheric phenomena* or similar terms. To simplify the matter for the readers of this book, I will stick with the word UFO. It rings a bell, so to speak, for most people who are interested in anomalous events.

Several recent developments should tell us something about what UFO researchers call the extraterrestrial (ET) hypothesis. For one thing, the number of reported UFO sightings has gone down significantly. Does this mean that the visitors from outer space have suddenly lost interest in our planet? In view of the countless UFO sightings during previous years, interpreted as visits to our planet, critics of the ET hypothesis have pointed out that it makes no sense to assume that a highly advanced civilization on some other planet would undertake such a ridiculously large number of trips through interplanetary space for whatever kind of purpose we might assume. Also, there are now far fewer daylight sightings. Most sightings occur in the middle of the night. Does this mean that the aliens piloting the extra-terrestrial spaceships have changed their travel plans? The shapes of the UFOs that are observed today are also different. When the wave of UFO sightings began after WW II, circular and rounded shapes were the most common. Now, many more triangular UFOs are observed than in previous years. Should we conclude that the aliens have had second thoughts about their engineering priorities?

The space travel notions associated with the UFO phenomenon increase the danger of being "taken for a ride" by authors who want to tickle the imagination and quench the thirst for sensationalism. More power to the debunkers in these cases. Here are some questions they have to deal with. I have reworded them for clarity. Let's look first at the question why alien beings would want to come to earth in extraterrestrial spacecraft. Is it because they (a) are curious about what's going on and just want to poke around, or (b) want to advance humanity and save us from self-destruction, or (c) want to prevent their own extinction by collecting samples of our genetic material, or (d) are conquerors who might attack us just for the hell of it? Secondly, there are questions about the actions of "the government". Did it devise some kind of cover-up because (a) it wanted to hide its dealings with aliens who landed on earth, or (b) it wanted to prevent a panic by suppressing evidence of the danger posed by alien invaders? If you anwered "none of the above", you are on the right track. All of the above, unfortunately, is the gist of what has been served up by authors for the above-mentioned reasons. But there's a flip side to all of this. It *is* a reasonable conclusion that the government used or fabricated reports of UFO sightings to hide some of its secret weapons programs behind a smokescreen of events that strengthened the widespread belief in UFOs. According to the newsmagazine *Der Spiegel*, a CIA report written in the early years of the cold war stated that 96 percent of the UFO sightings at that time were actually secret flights of spy satellites.(1)

It is possible that the CIA report was fabricated for cold war purposes. The CIA might have arranged for the document to be leaked, to disorient the communist enemy by exaggerating America's satellite capabilities. It is probably true that some of the UFO sightings were, indeed, misinterpreted observations of spy satellites. But based on the descriptions of UFOs published by researchers in more recent years, we can conclude that satellites and other manufactured flying objects cannot be an explanation for such a large percentage of UFO sightings.

Some of the UFO effects rule out this explanation. For one thing, the *luminance*, or optical power output, of many UFOs by far exceeds what engineers could produce for what would have to be flying power stations capable of accelerating instantly, as UFOs often do. Secondly, the flight patterns of UFOs *surpass the theoretical possibilities* of moving physical objects. UFOs not only accelerate on the spot to breakneck speeds. They execute precise right-angle turns while speeding along, then stop on a dime, or suddenly disappear from view.

The claim that most severely taxes the willingness of open minds to regard UFOs as an issue that should be taken seriously is the claim by observers of UFOs that they were abducted by alien beings. The self-described abductees often report that they observed their own body floating through the walls of their home into the interior of what they said was an extraterrestrial spacecraft. There, the aliens subjected the body to various procedures that have been described as part of the aliens' explorative undertakings on earth. Some of the reported procedures, such as the collection of sperms and ova, have obvious sexual implications. In their accounts of the events, the abductees say that they were unable to fend off the aliens because these possess the means to paralyze their victims.

Such experiences are very much like the trance experienced during hypnosis. I have already described the influence of magnetism in creating the sensation of *floating*. Part Two of this book explains how the anomalous events during a UFO-experience create abnormally strong magnetic forces. The sense of *paralysis* is like the hypnotic effect created by an overriding fear and the feeling of helplessness in the face of incomprehensible events. The sensation of floating without a recognizable cause contributes to this paralyzing emotional state. In support of claims that the encounters with aliens are real events, scars assumed to have resulted from surgical incisions performed by the aliens were occasionally reported, and objects described as implants inserted by the aliens were removed from the bodies of the abductees. The purported implants are misinterpretations at best, many are fabrications or hoaxes. They were found to consist of decidedly earthly material without a trace of what might be suggestive of transmitters designed by aliens to monitor human behavior.

Self-hypnosis could account for scars resulting from self-inflicted wounds, most of which are too superficial to qualify as evidence of the reported surgical intrusions. The self-hypnosis, in the case of UFO experiences, is not a voluntary act. It results from the processes that create the trance, and will be described in this book. Hypnosis can be a powerful influence on physiological processes. Experiments have shown that when participants allow themselves to be hypnotized, they develop a genuine blister on one of their fingers if they are told, under hypnosis, that a candle is burning underneath that finger.

A self-inflicted wound could be a fraud to "prove" that the reported event really happened, but it could also be an example of what is often encountered by psychologists investigating the subconscious motives and inner energies of troubled individuals. When going through a UFO experience, individuals *are* troubled. They are in a trance and experience normal sexual energies and desires, which accounts for the "explorations" by the aliens that involve sexual stimulation. Subconscious masochism could be a factor, which involves getting sexual pleasure from being dominated, mistreated, or hurt physically. Even if this is not the case, the powerful impression that what is happening is disturbingly true might create the subconscious urge to prove this "truth" to themselves by subjecting themselves to the procedures they are experiencing during the UFO trance. The experiences are reported as being very intense. "They felt like they were real, much more real than ordinary reality", is the way one abductee put it. (There's an explanation for this, and I will get to that in later chapters.)

All of the experiences reported by observers of UFOs, particularly those who say they have encountered aliens and were abducted by them, support the conclusion that these experiences are a complex a mix of fears, hidden expectations and suppressed desires comprising a broad spectrum of emotional and biological energy vibrantly alive in today's humanity. These psychological aspects of the UFO experience should not, however, obscure the obvious fact that the phenomenon involves physical forces such as anomalous luminance and abnormally strong magnetic effects. Equally compelling evidence that UFOs are a physical reality are the physical traces found at locations where UFOs reportedly landed and the blips on radar screens that correspond to the reported locations of UFOs in the atmosphere. I will describe the processes that create the physical and the psychological effects in Part Two of this book.

The magnetic effects are not only one of the causes of the abduction experience, but also account for the occurrence of anomalous physical events. In the presence of UFOs, as I have already said, compasses gyrate erratically and automobile engines stall. The abnormally strong magnetic fields generated by UFOs also explain why these "unidentified flying objects" are frequently observed in the vicinity of military installations and power stations. Adherents of the "extraterrestrial hypothesis" attribute this to the desire of visitors from outer space to explore the technical capabilities of our civilization. The actual reason why so many UFOs are observed at those locations is presumably the strong electromagnetic field created by the electronic equipment at military installations and by the generators of electricity at power stations. The interaction with the magnetic fields of UFOs creates a force of attraction. The traces on radar screens cited as evidence for the alleged exploration of locations on earth, moreover, are only chaotic patterns of blips that provide no indication of flight patterns guided by a rational intent.

*Crop circles* and *cattle mutilations* are also often reported in the context of UFO phenomena. Crop circles are circular patterns of flattened stalks in fully grown fields that are about to be harvested. The patterns are geometrically perfect and appear overnight in areas where UFOs were reportedly seen. Some are not just simple circles, but complex configurations. Cattle mutilations usually occur, unobserved, in remote areas. Local farmers have said that they saw a UFO as well as unmarked helicopters in the area where they later found the bodies of the mutilated animals. Various parts of the bodies had been cut off. The incisions were razor-sharp. Often, there was not a trace of blood.

Such stories, obviously, lend themselves to the type of attention-grabbing sensationalism that I described earlier. They often have little or nothing to do with the facts established through scientific investigation of paranormal events. Instead, they reinforce the mind-sets and expectations created by publications "selling" their viewpoints. The dominant ones are the belief in nuts-and-bolts spaceships used by non-human space travelers and the opposite argument that UFO phenomena are in large part government cover-ups (smokescreens) to hide secret weapons research.

The smokescreen argument is preferably used by those who wholeheartedly pursue the purpose of debunking the whole "ufology" craze. That craze, paradoxically, also includes a cover-up scenario. As already mentioned, some ufologists argue that the awesome powers of the alien intruders cannot be officially acknowledged because this might cause a panic, or that the government is secretly cooperating with the aliens to develop space technology.

The crop circles, according to one of the smokescreen interpretations, are created in military tests of high-flying platforms emitting laser-guided rays. The cattle mutilations are attributed to tests of remote-controlled, high-precision instruments that can be used as weapons. The black, unmarked helicopters reportedly seen where the mutilations occurred, on the other hand, have been interpreted as evidence that "the government" is perpetrating the mutilations by ordinary means just to keep the belief in UFOs alive, to obscure a variety of secret weapons research programs.

Among the ufologists who regard the mysterious events as the activities of aliens visiting earth, some argue that the aliens create the crop circles to alert mankind to their presence and to demonstrate their intellectual and technical capabilities. Mutilations of farm animals and the removal of body parts, according to this scenario, occur because the aliens are investigating the life forms on earth.

The extraterrestrial (space travel) interpretation of the UFO phenomenon as described above has been largely abandoned by serious researchers. One of the down-to-earth explanations of the mutilations is that they are the work of satanic cults. Some police files suggest that this might be the answer. Many of the crop circles are cases of hoax and fraud, but with others it is hard to imagine how the geometric precision of the complex patterns could have been hastily and silently created overnight. Besides, there are magnetic effects in crop circles, as well. The batteries of the equipment carried by investigators often fail, and when tractors are driven into the mysterious crop formations, their engines stall. The fact that in many instances the flattened stalks were not broken, but continued to grow, points to anomalous biological effects. These effects, and the anomalous atmospheric phenomena that create the patterns of the crop formations will also be described in Part Two of this book.

UFO researchers who have abandoned the theory that aliens come to earth in spaceships have turned their attention to theories of complex physical processes. What might be capable of traveling through space, they argue, are processes associated with mind events that overcome the limitations of time and space. This is in keeping with the scenario proposed by the Physics of Encounter. Such processes generate a variety of physical effects that influence the events inside atoms. Inner-atomic events affect the experiences generated by the atoms of the brain and can alter biological processes.

So much for this overview of the UFO phenomenon. The Physics of Encounter leaves no doubt that the phenomenon must be studied in the context of anomalous mind events. In this context, several other concepts have also created distinct unease in the scientific establishment. The concept of *precognition*, for example, contradicts the traditional interpretation of what seemed to be an ironclad law of physics: the law of cause and effect. Precognition allows us, on unpredictable occasions, to foresee what will happen. The most frequently reported instances of this occur in precognitive dreams. In some well-documented cases, the event seen in a dream actually happened later, with many details corresponding precisely to how the dream was described to others before the event took place. A precognitive or prophetic dream transforms the *possibility* of a future event into experienced *reality*. The experience, in this case, first occurs in a dream and later becomes the experience of an actual event.

The possibility of an event, as explained, is contained as a virtual effect in probability waves. That's standard vocabulary for theoretical physicists. Nothing about the future is absolutely *certain*. The occurrence of an event has varying degrees of probability. In some cases, the probability is so high that it is the practical equivalent of certainty. But the theory of quantum mechanics states that all events that become observable as physical reality arise from absolutely random, haphazard quantum events. In theory, the coffee you are about to drink could suddenly pop out of your cup, the way a lone particle pops into existence in the world of quantum events. But for the above-mentioned event to occur, all of the countless quantum events occurring at random in your cup of coffee would have to occur in unison, in one direction. The likelihood of such an event is so low that you need not worry about it.

We have to differentiate between subatomic processes that create mind events and the much more encompassing processes that add up to observable matter events. In the scenario proposed by the Physics of Encounter, both processes involve encounters between the expanding VE-surfaces that are probability waves. The encounters occur in morphic resonance. A dream, obviously, does not have to come true. The dream represents a possibility. It is a *micro-event* in the brain that does not necessarily correspond to a *macro-event* involving "real things", an interaction between solid objects. Micro-events occur when particles suddenly materialize in space, seemingly out of nowhere. The particles do not predictably appear at a particular location, at a particular time. They are suddenly "there". The materialization is a matter of chance, or uncertainty, following the probability laws of quantum mechanics. Micro-events, as I explained, occur when probability waves, which are the expanding surfaces of VE-spheres, resonate with one another at points of encounter, i.e. if they interact in morphic resonance.

Precognitive morphic resonance can result in the *intervention paradox*. If you dream that something awful will happen, you might decide to do something to prevent it, just in case the dream is true. If it was indeed a prophetic dream and you succeed in changing the course of events, you thereby falsify the precognition that was so helpful to begin with. What you foresaw does not happen after all. Let's say you dreamed that a blue car will veer out control at a certain location and that a friend who always walks by there is killed. You warn the friend to avoid the location. The accident involving a blue car does indeed happen, at just about at the time your friend would have been there on his daily walk. But thanks to your intervention, you could assume, your friend is alive.

Did your intervention really falsify the dream? What you saw in your mind was a *possible* event, not a certainty. Some of it actually happened. Maybe your friend would not have been killed, even without your warning. There is no way of knowing. The probability waves that reached your brain, originating in the future, deposited some aspects of a possibility in your brain because they were meaningful to you. In the example I made up, the blue car could be associated with some powerful memories involving a car more or less like that, and the location of the accident could be meaningful because you often accompanied your friend there. Moreover, something may have occurred before your dream that made you fear you would lose your friend. All of this contributed to the morphic resonance that caused you to see a specific event in your dream.

The term "precognition", as this hypothetical example shows, is too broad. It means *knowing* of something in advance. But as I said, some of the main aspects of what is "foreseen" may not occur, even without an intervention. Nonetheless, it is a fact that a future event, a macro-event involving solid objects, can influence the "here and now". In my example, a blue car did cause an accident at that particular location and that future event influenced your dream. But the future event was not a certainty until it actually happened. The future is not predetermined. It is filled with countless possibilities, some more probable than others. In an act of free will, the driver might have chosen a different route that day. Or, at the last moment, he might have noticed that he was being inattentive at the wheel and avoided the accident.

The Physics of Encounter suggests that the influence of a macro-event, backward in time (in this case, from the future to the present), is the sum of the micro-events that occur in the material substance that is involved. This applies to the inanimate matter of the car in my example, including the electromagnetic waves that identify its color, and also applies to the biological matter of the driver's brain. The micro-events are a product of the interaction between the surfaces of VE-spheres. The influence of the vacuum energy in these surfaces is transmitted instantly across space and time. All aspects of the future exist in the present moment - as possibilities.

The pivotal concept in this scenario is the concept of *time*. In the pre-manifest reality of quantum vacuum fluctuations, vacuum energy fluctuates as VE-surfaces expand and collapse, due to encounters that occur in morphic resonance. These events occur without the elapse of time. Time, as Einstein explained, is "relative". The Physics of Encounter suggests that time is a process energized by the vacuum energy that energizes the creation of mind events. Where observers are present, time "flows", relative to the location and motion of the observer. Where no human observer is present, the elementary, pre-manifest physical reality in our universe is an eternal NOW.

The timeless elementary reality is intertwined with the reality of our everyday experience. This is the *entanglement* that plays such an important role in the theories of paranormal phenomena. At this point, I think it will be helpful if I introduce a metaphorical image to clarify the process through which the minds of people living today can influence, or be influenced by, events that have not yet occurred. Where in space is that timeless reality located that the mind can "pick up" through precognition? For the sake of the argument (the metaphor), I will assume that you drink coffee. You need water to brew coffee. When the coffee is in your cup, it is the result of a mix between two things: water and miniature bits of a solid substance, the ground coffee beans. The water has picked up the *taste* of the substance. Enjoy your coffee! You do not have to think about the fact that your experience is based on two different kinds of reality. When you experience the effects of one reality, you also experience the effects of the other. The water is needed for what you are experiencing, but it's coffee you are drinking, not water.

The effects that create the taste of coffee are inseparably mixed with the reality of the water. The taste of coffee that you enjoy during your breakfast is "in" the liquid you drink, but only as a *possible effect*. The effect does not occur if you do not drink the coffee. The possibility of the taste is the timeless "other reality". The unobservable reality from which nature brews the "flavors of life" (the taste of coffee and all other qualia of consciousness) also creates the sense of touch. It creates the quantum mechanical events that are the substance of physical objects. What exists as physical reality and what you experience originates in the quantum vacuum fluctuations of the universe.

All events, all experiences, arise from the quantum vacuum fluctuations through which, as Hal Puthoff pointed out, we are in touch with the entire universe. Puthoff described these fluctuations as a "blank matrix upon which coherent patterns can be written", providing a broad spectrum of information structures, "possibly even non-biochemical components of memory". The fluctuating vacuum energy, Puthoff said, corresponds in large part to the pre-scientific images of an "all-pervasive sea of energy that undergirds (...) all phenomena, interconnecting mankind and cosmic reality". Puthoff's conclusion: a synthesis of such images and modern scientific concepts is within reach.(2)

The arguments of the physiologist John C. Eccles, who was awarded a Nobel Prize in 1963, support the scenario that the entanglement between the two kinds of realities in our universe is a two-directional flow of influences. Eccles, like other scientists, pointed to the distinction between measurable *brain events* and non-measurable *mind events*. Eccles, however, went beyond conventional wisdom and described the mind as an "independent reality" that acts upon the brain "in accord with its attention and its interests and integrates its selection to give the unity of consciousness from moment to moment".(3 )

Eccles used the term "self-conscious mind" to describe a mind that is aware of its own existence. The mind, in his view, is more than just "neural machinery" (brain activity involving neurons, or nerve cells). It is more than just a computer that processes information. Eccles identified the relevant portions of the brain as a "cerebral hemisphere" and pointed out that "the unity of conscious experience is provided by the self-conscious mind and not by the neural machinery of (...) the cerebral hemisphere. (...) For example, when following up a line of thought or trying to recapture a memory, it is proposed that the self-conscious mind is actively engaged in searching and in probing through specially selected zones of the neural machinery and so is able to deflect and mould the dynamic patterned activities in accord with its desire or interest. (...) This intervention of the self-conscious mind upon the operations of the neural machinery is exhibited in its ability to bring about movements in accord with some voluntary desired action (...)."(4)

The mind/matter model proposed by Eccles leaves one question open: how does the mind manage to maintain its independence from the brain? From where does the mind get the energy to influence the brain, instead of vice versa, as assumed by mainstream scientists? The Physics of Encounter suggests that a mind that is anchored in the brain can maintain a limited independence because it is energized by the fluctuating vacuum energy pervading space.

The two entangled types of space are what I will call the space of vacuum energy (*VE-space*) and the quantum mechanical space (*QM-space*) in which the "neural machinery" of our brain allows us to experience the space in which we observe physical objects. For the human mind, both types of space are sources of information about "outside reality".

In the Physics of Encounter, *all* mind/matter interactions, not just anomalous events, involve an entanglement between QM-space and VE-space. The taste of your coffee results from such an entanglement. It is a perfectly normal effect that combines the measurable physical reality of the liquid you are drinking (coffee) with the non-measurable reality of the taste that is contained in the liquid as a potential effect that can influence your mind.

In paranormal phenomena, the entanglement deviates from the usual mix. That happens when morphic resonance allows an unusually large amount of the *vacuum energy* from VE-space to kick in. There is no clear-cut dividing line between a normal and an anomalous entanglement. If the input of effects originating in VE-space reaches a certain critical level, however, the process begins to speed up on its own. At a certain point of cumulative *quantitative* changes, a change in *quality* suddenly sets in. The literature on dialectical materialism abounds with examples of such processes. They create anomalous events when morphic resonance strongly and continuously favors the selection of certain types of *superpositions* (information structures) in the wave patterns of interacting VE-surfaces, thereby crowding out normal perceptual information. The changed quality of mind events can be a trance or what researchers of paranormal phenomena call an *altered state of consciousness*.

The influx into the human mind of information structures encoded in the fluctuating vacuum energy of space can also result in the anomalous acquisition of knowledge or capabilities. As Puthoff said (7/2), quantum vacuum fluctuations are "a blank matrix upon which coherent patterns can be written". The information structures, according to Puthoff, might include "non-biochemical components of memory". In other words: the bio-chemical events in the nerve cells of a person's brain might not be the only repository of that person's memories. This highlights the urgent need to investigate the cases of anomalous mind events that are experienced as memories of a previous life and are interpreted as cases of *reincarnation*.

Ian Stevenson, who was chairman of the Department of Psychiatry and Neurology at the University of Virginia, recognized that need and collected hundreds of cases (more than any other investigator) during travels that took him to India, Ceylon, Brazil, Alaska, and Lebanon. Stevenson's work was reviewed in the *Journal of the American Medical Association*, and the reviewer's conclusion was that the evidence regarding reincarnation "is difficult to explain on any other grounds". The reviewer of the *American Journal of Psychiatry* wrote that the cases were "recorded in such full detail as to persuade the open mind that reincarnation is a tenable hypothesis to explain them."(5)

My view is that the concept of reincarnation, as it is commonly understood, is too diffuse to be scientifically helpful. It merely makes clear that, yes, here is something that calls for a description of the underlying processes. For one thing, if we apply the concept of reincarnation to the American woman who experienced what she felt were memories of her previous life as Napoleon, as described by Almeder (6/1), then why is it that only some people are aware of this kind of "reincarnation"? According to religious interpretations of that concept in other cultures, every living person is a reincarnation of some other person (or animal). With the billions of humans on this planet rapidly approaching the two-digit mark, how can the many human minds of later incarnations be matched up with the few humans who lived in the earlier stages of human history and evolution? Would any humans alive today experience themselves as creatures from the Neanderthal era? How meaningful is it to ascribe the mind of an ant or a zebra to a reincarnated human?

If we apply the concepts of timeless processes and of morphic resonance to the above spectrum of anomalous phenomena, we can resolve the logical dilemma. Many psychologists and investigators have used hypnosis to guide human minds back in time to buried memories, and some investigators experimented with this *hypnotic regression* to uncover buried memories of what their subjects described as a previous life. Others have used *hypnotic progression* to take the minds of their subjects into the future. Chet Snow is one of the investigators who enabled his subjects to describe what they experienced as their *future incarnations*.(6) Hypnosis, in the scenario proposed here, creates an entanglement between the world of timeless processes and the world of physical reality that creates the experienced flow of time. The entanglement occurs in morphic resonance between the VE-surfaces, or probability waves, that connect the two worlds in our universe.

VE-surfaces interacting in morphic resonance, as I explained, create both mind events and observable physical reality. That is why the bodies of people who are "reincarnated" often have birthmarks or congenital deformities corresponding to those of the person who is considered to be the previous personality.

A final conceptual difficulty that I will address here is the distinction between what physicists have called the *agent* and the *receiver* of anomalous influences. The difficulty arises because the events that create anomalous effects are instantaneous events. They occur without the elapse of time. Moreover, these events create two-directional effects. In the imagery used by Eccles (7/3), the "independent" reality of the mind influences the "neural machinery" of the brain, and the events produced by this machinery, in turn, influence the mind.

The difficult distinction between agent and receiver, as the physicist Garret Moddel explained, often makes it "difficult to distinguish between *precognition* and *psychokinesis*. In psychokinesis, the agent produces a reaction in a receiver. For example, if one chooses to affect the output of a random event generator, the agent is the person and the receiver is the random event generator." It can also be the other way around. In precognition, Moddel pointed out, a person "may be convinced that he is sensing and not causing a particular future event, i.e. that he is the receiver and not the agent." But that may not be so. "The act of sensing an event may well be one and the same as the act of influencing it. By analogy with quantum mechanics, the act of measuring a phenomenon also affects it."(7)

The laws of physics make it quite clear that the energy needed to make something happen as a measurable physical reality does not just appear like a rabbit out of a hat. In the Physics of Encounter, the energy that creates anomalous events is not (yet) "real", measurable energy, but hidden energy. The phenomenon manifests itself in what Moddel called the "entanglement of precognition and psychokinesis". It is "tempting to speculate", he said, that the entanglement involves processes like those that occur in quantum mechanics, "in which the act of receiving information from a system also influences that system." Moddel pointed out that the anomalous influence "must act when it can have an effect without expending energy. This can be accomplished by influencing the outcome of an ostensibly random event (....)." Random, unpredictable quantum events, Moddel continued, are microscopic events. "To produce large, macroscopic effects, there must be an amplification of the quantum trigger, or there must be a large number of triggered events occurring in tandem."(8)

The "tandem" occurrences, as interpreted here, correspond to the co-occurrences described by Manthey (4/9). The "quantum triggers" are encounters between the surfaces of VE-spheres containing the quantum vacuum of QV-bubbles. The amplification of the vacuum energy in the trigger mechanism occurs when abnormally large numbers of encounters between VE-surfaces occur in morphic resonance. Since each encounter is a random event, the probability of anomalous events releasing large amounts of measurable energy is correspondingly low.

There is a good measure of dialectic paradoxes in the above scenario. It describes how unobservable events can have an influence "without expending energy". It describes how an unobservable cause can release large amounts of measurable energy. It describes how a quantum event is created by interacting QV-bubbles, of which each is a quantum vacuum. It describes a process during which, in a single instant, the human mind is both an agent and a receiver.  The paradox is resolved if we understand that a QV-bubble contains the elementary ingredients of consciousness (qualia).

In the world of timeless events that is an integral part of this scenario, there is no clear-cut boundary between the "self" (the conscious mind) and the non-self reality of the universe.  Both the conscious mind and the universe are an agent and a receiver.  A theory presented by John Wheeler clarifies the concept of a dual reality that comprises both consciousness and matter, both timeless reality and time.  Wheeler argued that acts of observation, as quantum events, can cause other quantum events in the distant future or the distant past.  The human mind, in Wheeler's theory, may or may not be involved in the paradox of quantum events.  As explained by Tim Folger in the science magazine *Discover*, Wheeler said that a particle of matter "exists in many possible states at once (...and is...) not quite real and solid until it interacts with something (...).  When that happens, one of the many different probable outcomes becomes real."  An object that is "not a conscious being (...) transforms what might happen into what does happen (...).  Wheeler suspects that most of the universe consists of huge clouds of uncertainty that have not yet interacted either with a conscious observer or even with some lump of inanimate matter.  He sees the universe as a vast arena containing realms where the past is not yet fixed."**(9)**

In Wheeler's theory, the past as well as the future is an open book.  Tim Folger explained that, to Wheeler, "we are not simply bystanders on a cosmic stage; we are shapers and creators living in a participatory universe.  Wheeler's hunch is that the universe is built like an enormous feedback loop, a loop in which we contribute to the ongoing creation of not just the present and the future but to the past as well. (...) We are tiny patches of the universe looking at itself – and building itself. (...) By peering back into time, even all the way back to the Big Bang, our present observations select one out of many possible quantum histories of the universe."**(10)**

> *Ah, time is a riddling thing,*
> *and hard it is to expound its essence!*
> Thomas Mann

## NOTES

(1) *Der Spiegel* (1997), p. 146
(2) Puthoff (2002), *personal communication*
(3) Popper and Eccles (1977), p. 355
(4) Ibid, pp. 362-364
(5) Tucker (2008, p. 38
(6) Snow (1991), pp. 1-269
(7) Moddel (2004), p. 297
(8) Ibid, pp. 298-299
(9) Folger (2002), p. 47
(10) Ibid, pp. 45 and 47

# *Chapter 8*

**UFOs and consciousness fields. Imposing structure on random events. Playing with a whimsical robot. Resonance, Near-Death-Experiences, and non-local proximity.**

This chapter will take up John Wheeler's image of a "universe looking at itself", with the human mind participating, as described at the end of the preceding chapter. The scenario is in keeping with the arguments of John Eccles (7/3) that the mind is an independent reality which influences the "neural machinery" of the brain, while these matter events, conversely, influence what happens in the mind. Equally relevant in this context is Garret Modell's argument (7/7) that mind and matter are, at the same instant, "agents" as well as "receivers" of influences on each other.

The relevance of these insights to the study of anomalous events is most readily apparent by looking at the specifics of the UFO phenomenon. The Physics of Encounter suggests, based on the above reasoning, that the phenomenon is a mind/matter entanglement in which the observation of a UFO is not only a mind event arising from events in the brain, but also a mind event that occurs outside the brain, in the atmosphere where the UFO is located. At both locations, the mind events are part of a larger whole. The UFO in the atmosphere corresponds to what Moddel called an "agent" of an influence, and this influence impacts on the receiver, which is the brain of the observer. The converse is equally true: it is the brain that triggers the creation of the UFO.

The anomalous phenomenon, in this scenario, is not only evidence that mind events can occur outside the brain. It also shows that the mind events of many humans, occurring in morphic resonance, can act as a composite trigger that releases the energy needed to create what is observed as an "unidentified flying object".

UFOs are a social and cultural phenomenon that illustrates the influence of what investigators of paranormal events have called the extended mind. The experiments done by Schwartz and Russek (5/10) have shown that merely *imagining* to be staring at someone can influence the other person's brain. The influence on the atoms of the other person's brain is, in essence, the creation of quantum mechanical events in those atoms, allowing the "receiver" to become aware of what the "agent" is imagining. When a UFO is observed, the receiver consists of the quantum mechanical events in the atoms of the "anomalous atmospheric phenomenon" (the UFO) that is created by the collective imagination of many people.

Before the appearance of a UFO, the quantum mechanical events in the atoms of the atmosphere are normal random events. Nothing extraordinary occurs. Storms, thunder, and lightning are normal events arising from the interactions of forces in the macroscopic world, and the macroscopic world, in turn, is built from the random interactions in the microscopic world of elementary particles. But the effects of the mind can instantly impose structure, or coherence, on random events. The "Global Consciousness Project" founded by the psychologist Roger Nelson supports this scenario. As reported by Michael Lemonick in *Time* magazine, the project is based on the theory that electronic devices installed at locations around the world "may be picking up some sort of planetwide field of consciousness".**(1)** The project uses random-number generators. According to Nelson, world events that cause strong emotions are correlated with a "persistent departure from what is expected of random data".**(2)**

The effect that created what Nelson called a "detectable non-random structure" can be visualized as the equivalent of flipping a coin and finding that, say, "heads" consistently come up much more frequently than "tails". In the experiment, still ongoing, an unobservable reality corresponding to a world-wide arousal of emotion surges through space and impacts on about 60 devices located all over the surface of the earth.

The events that registered on the devices were powerful attention-grabbers such as the funerals of Princess Diana and Mother Teresa, the fall of the Berlin wall signaling the demise of Soviet-style communism, international sports events such as World Cup Soccer or the Olympic Games, and widely observed holidays. The evidence, Nelson said, suggests the existence of a "communal, shared mind in which we are participants (...)." It is a "consciousness field" in which large numbers of people become "deeply absorbed in one focus". The focus is created by events that "inspire strong coherence of attention and feeling".**(3)**

The "non-random structures" of UFOs owe their existence to precisely this kind of emotional focus in a "consciousness field" that encompasses many shared mind events. The focus, in the case of UFOs, involves not only shared emotions. It is also a *conceptual focus*. The physical event in the sky interpreted as a UFO usually originates at the location of an atmospheric disturbance, or some other misinterpreted event in the sky. It arouses strong emotions that are a mix of fear and intellectual excitement because the observer does not recognize the natural cause of the phenomenon. Emotionally charged questions surge through the observer's mind. Could this be an extraterrestrial spacecraft? The present era of human history, marked by the exciting onset of space travel and the titillation of science fiction tales, predisposes many people to such a conclusion. Even before television began to broadcast mankind's adventures in space, minds were flooded with fantasies of space travel. The imagination was energized by books, magazines and media reports, and by images bountifully provided by the entertainment industry. Intergalactic battles raged in movie theaters and on television screens.

The consciousness field is a field of hidden energy, an unobservable, non-measurable physical reality. It consists of countless extended minds influencing each other in morphic resonance, and influencing the events in space. It is a field of fluctuating vacuum energy that contains not only the elementary constituents of consciousness, but also memories that are buried in the subconscious mind. As Puthoff pointed out, the patterns encoded in the quantum vacuum fluctuations of space provide the mind with information structures. The Physics of Encounter suggests that the field of hidden energy corresponds to what the psychiatrist Carl Gustav Jung called the *collective unconscious*.

The collective unconscious includes not only the mind events occurring today, but also all events experienced by our distant forebears, including their fears, thoughts, religious images and interpretations of reality that preceded our scientific insights. Also included are the experiences and greater knowledge of the generations that will come after us. In sum, the collective unconscious is a volatile mix of future wisdom and past misconceptions, of rational thought and emotions.

In all cultures throughout recorded history, evidenced in myths, legends, and folklore, the effects of the collective unconscious have created strikingly consistent images that the psychiatrist C. G. Jung called *archetypes*. They are, according to Jung, ideas or perceptual patterns ingrained in every human mind and expressed in dreams or in art, and in certain symbols common to all mankind. As interpreted here, archetypes arise in the human mind when it encounters events perceived as existential challenges, events it cannot understand and that seem threatening. They exist as encoded perceptual patterns in the collective unconscious.

During millions of years of evolution, when humans had not completely emerged from the animalistic stage, the human mind encountered largely unknowable other beings as animals. The animalistic archetypes have survived in the collective unconscious and surface when normal minds succumb to the paralyzing fear of the unknown. The "aliens" emerging from UFOs, therefore, may appear in (partially) animalistic form. Today's UFO phenomenon arises from the expectations created in our age of space travel. Much of it mirrors the crass sensationalism of media reports. Accordingly, the aliens usually look like "extraterrestrial", human-like creatures.

In highly industrialized countries, the media are a powerful influence. They shape images and beliefs. Why is it, then, that so many UFOs are reportedly observed in other countries, in Africa, in China, and in Latin America, for example, and even in the Antarctic? In Russia, scientists investigated the phenomenon while that country was still governed by communists. UFOs were frequently reported, although the government-controlled media avoided the UFO hype of western countries. The answer suggested here is that the contents of the collective unconscious most recently created in the industrialized countries are readily picked up worldwide, through morphic resonance.

All of the experiences reported by observers of UFOs, particularly those who say they have encountered aliens and were abducted by them, support the conclusion that these experiences are, as I said, a complex a mix of fears, hidden expectations and suppressed desires comprising a broad spectrum of mind events in today's humanity.

>*Let us assume that the abduction experience is an extraordinary*
>*type of dream.*
>*The coherence of the experience shows it is not a private,*
>*but a collective dream.*
>*A dream, perhaps, of the species mind.*
>*Produced by the species mind, like any dream,*
>*it is about the dreamer.*
>*Perhaps about the dreamer's future.*
>Michael Grosso

When alien beings communicate with those they have abducted, it is invariably about the future of mankind. They urge their captives to recognize that this future is endangered by the follies of the human species. These wordless conversations are, of course, occurrences in a troubled human mind.

If we allow the assumption that such anomalous mind events are a mix of many experiences that have characterized the human mind since it came into existence, we arrive at the model of mind/matter reality proposed here. UFOs are apparitions that arise from an interaction between the collective unconscious and the physical reality underlying our consciousness. The physicist Jaques Vallee has documented that UFOs and aliens are phenomena comparable, in their psychological origins, to apparitions throughout the ages.(4) In the legends of Native Americans, for example, immortal beings emerged from canoes that descended from the heavens. In both meanings of the word, the appearance of beings from other worlds, and the appearance of their means of travel, correspond to the dominant expectations of culturally conditioned minds.

The mysterious airships seen over parts of the USA in 1897 were the UFOs of that time. They mirrored the excitement about a newly created industry that was introducing gas-filled, lighter-than-air dirigibles, but they also seemed to be a portent of future developments. Newspapers were filled with speculations about secret inventors, because the performance of the mysterious airships surpassed all known technological capabilities. Drawings of the unidentified airships published in newspapers show that the structural details, including the configuration of position lights, corresponded to those of a type of airship that was built much later, in 1911. As the French UFO investigator Fabrice Bonvin pointed out, the airships of 1897 did not yet have position lights.(5) In that sense, the anomalous events that were the UFOs of 1897 contained information about future developments. Part Two of this book provides details about the anomalous transfer of information across the barriers of time.

In the Second World War, American Air Force pilots occasionally reported seeing small, mysterious aircraft that avoided closer engagements and were only about five feet in diameter. They called them "foo fighters". The unexplained sightings stopped after the war. Today's UFOs are usually between thirty and ninety feet in diameter. The modern era of UFOs began in June 1947, when Kenneth Arnold, a seasoned pilot specializing in search and rescue missions, reported that he had seen nine disc-shaped objects flying in formation over the mountains of Washington State. He said that they looked like saucers and flew as if they were skipping over a surface of water. When people began seeing similar objects in the sky, they called them "flying saucers". Later, unidentified flying objects of other shapes began appearing. The name UFO came into general use.

The probability that paranormal mind events will occur is increased if people in the same cultural environment absorb an unusually large amount of emotionally charged information. The result is comparable to a religious frenzy or rampant superstition. In the minds of many people, the "aliens" arriving in spaceships from a distant galaxy are messengers of a super-human power that can rescue our world from the brink of disaster. Sociologists have likened the UFO craze to a pseudo-religion.

UFO events witnessed by many people can be triggered by a misinterpreted event in the sky observed by only *a few individuals* whose mindset favors such a startling perception. The phenomenon occurs in a cultural setting that is comparable to a *super-saturated liquid*. That kind of liquid is created through the gradual absorption of, say, grains of salt. The grains dissolve in the liquid and begin to saturate it. At some point, dropping one tiny additional grain of salt into a glass filled with such a liquid, and/or simply perturbing it, will produce an astonishing result: almost instantly, the saturated liquid becomes a solid block of salt. The substance that is culturally absorbed, in this metaphorical example, consists of esoteric imagery and fantasies that have little to do with the measurable physical effects of UFOs.

One argument made against the physical reality of UFOs is that observers of the same event, at the same location, often disagree on what they have seen. The reported shape and other characteristics of the observed object vary. In some cases, a UFO observed by a large number of people is not seen at all by others in the same area. Between 1989 and 1991, about 3000 people in Belgium reported that they saw UFOs. While they watched huge triangular objects gliding silently over heavily populated areas, others saw nothing. In the area of Gulf Breeze, Florida, between November 1987 and May 1988, about 135 people said they saw UFOs. Many of them took photos, about 60 in all. One observer reported that the UFO on the photo he took looked different from the UFO he actually saw.

The explanation suggested here is that the "appearance" of a UFO, in both meanings of the word, is a random event that depends on a specific state of mind. The minds of different people pick up different aspects of the collective unconscious. Depending on their momentary predisposition to observe what is happening around them, the collective unconscious may not have any influence on them at all.

One consequence of the over-saturation with UFO-related information is the appearance of what researchers describe as *Men-in-Black*. According to the people who report having encountered them, the Men-in-Black approach individuals who have observed a UFO and sternly tell them not to talk about the observation. This imagery mirrors a widely held belief that the government is engaged in sinister efforts, including the intimidation of witnesses, to control the flow of information about UFO events in which it is involved. According to some accounts, the government has persuaded the alien occupants of UFOs to cooperate with scientists working in secret research facilities. Stories of government cover-ups are grist for the mills of less than reputable news media, magazines, books and the entertainment industry.

The Men-in-Black are reportedly seen arriving or departing in automobiles of by-gone times. They wear black suits that are no longer modern, walk stiffly and have a mask-like facial expression. They seem like distorted images of evil-minded characters in old movies. The experiences of being approached by Men-in-Black occur in a partial trance. The people who see them may be fully aware of their surroundings. In one case reported by Bonvin, a Man-in-Black intruded on a UFO conference that Bonvin was attending, stepped up close, sternly spoke his warning while Bonvin was following the proceedings, and then dissolved.(6) Such mind events illustrate what I explained in chapter seven: the boundary between a normal experience and a paranormal experience is blurred. The degree of "*entanglement*" can vary.

The proposed model of anomalous processes associated with the collective unconscious has much in common with what Bonvin has called an intraterrestrial interpretation of the UFO phenomenon.(7) This interpretation favors the so-called *GAIA hypothesis*, which assumes that the earth and everything it has brought forth is like a conscious super-organism defending itself against the dangers resulting from the activities of humans, which threaten the environment and the continued existence of mankind through weapons of mass destruction. The GAIA mind is very much like the collective unconscious as interpreted here. The two models of mind/matter reality differ mainly with regard to UFO events that are clearly erratic or nonsensical and serve no recognizable purpose. Bonvin acknowledged a multitude of such inconsistencies but argued that the quirky aspects of the UFO phenomenon serve a rational purpose. They are, in his view, today's counterpart of the legendary silly games played by goblins and imps enticing humans to watch and to become involved. The GAIA hypothesis suggests that what seem to be pranks are a part of an overarching purpose to attract attention. In that sense UFOs, like the apparitions of former eras, are instruments of the GAIA mind.

Is the mind of a GAIA super-organism responsible for the creation of crop circles in the fields that are the symbolic source of our daily bread, and for the spectacular killing of cattle, the symbolic source of meat and milk? Is that its way of making us aware that we are jeopardizing the natural basis that sustains our lives, as Bonvin suggests? Or should we attribute the phenomena to physical forces arising from the collective unconscious? The GAIA hypothesis cannot explain away all of its inconsistencies. If the flights of UFOs are undertaken to draw attention to sites of irresponsible and dangerous activities, such as military installations harboring atomic missiles, why are there no massive sightings of UFOs over areas where rain forests are recklessly destroyed, where oil slicks and poisonous waste pollute oceans and rivers?

In view of the above, the most reasonable assumption is that UFOs and other anomalous phenomena are the product of hidden energy that corresponds to Nelson's field of global consciousness and to Jung's collective unconscious. This energy is contained in the quantum vacuum fluctuations of space through which, as Puthoff argued (7/2), we are in touch with the entire universe.

The observation of UFOs produces psychological effects that are identical with the effects produced by experiencing *real* events. The late Harvard University psychiatrist John E. Mack examined a large number of people who reported that they experienced what seemed like an abduction by alien beings. Mack came to the conclusion that an *outside influence* on perfectly normal minds causes such experiences. They often result in post-traumatic stress disorder (PTSD). The symptoms of PTSD caused by an abduction experience are the same as those of accident victims or battle-shocked soldiers. PTSD can also be caused by an *hallucination*, an imagined event that was experienced as disturbingly real. But PTSD symptoms caused by abduction experiences are clearly different from those caused by hallucinations. Post-traumatic stress disorders create measurable irregularities involving the nerves and the skin, and the irregularities caused by abduction experiences are precisely like those caused by real events, not like those caused by events that were simply imagined.

The Massachusetts Institute of Technology held a conference on abduction and related issues in 1992, focusing on the question how the science of physics can contribute to a clarification of such anomalous, mind-related phenomena. A magnetic force acting on the brain, as I explained, can create the experience of being floated away against one's will. The second part of this book contains the proposed scenarios of the processes that create physical effects (e.g. magnetism, electric charge, and mass) where these effects would not normally occur, or where they are abnormally strong. The conference at MIT is one of the heartening instances showing that this subject matter is taken seriously at the highest levels of scientific research.

A large variety of experiments support the conclusion that the anomalous effects of mind events on physical reality may be attributable to a field of hidden energy involving subconscious desires, amplified through morphic resonance with information structures in the collective unconscious. A striking example of this is the *experimenter effect* that seems to have a vexing influence on experiments testing the statistical probability of an anomalous event. If scientists are confident about their theories and expect a certain outcome of their experiment, that experiment is more likely to produce those results. This goes both ways. A skeptical attitude can prevent an outcome that would otherwise occur. In a séance, when like-minded participants assemble around a table and attempt to produce anomalous forces causing objects to move, for example, one skeptical onlooker can spoil the whole thing. When, in the successful healing experiments with mice conducted by Moga and Bengstson (6/14), the laboratory assistants who had been chosen at random were replaced by biology students, the results were much less impressive. The experimenters attributed this to the skeptical, scientific attitude of the students.

The Physics of Encounter suggests that the human mind, through morphic resonance with other minds, can tap into a reservoir of hidden energies. Access to the reservoir may be facilitated by the surge and the focus of energy created by the shared mind events of participants in séances, for example. In all cases, the effects are amplified by morphic resonance with the collective unconscious. The latter contains the hidden energy and information structures of countless mind events associated with all humans in the history of mankind.

If the reports about séances are true, the willful intentions of participants assembled around a table may not only cause the table to rise and to tilt, but may also trigger loud raps on the table, interpreted as indications that the disembodied spirit of someone who is no longer alive is communicating with the group. The concept of a "bodiless person", described by Almeder (6/2) in a different context, is in keeping with the scenario that the collective unconscious contains the information structures created in the minds of persons while they were alive.

To test whether the combined mind effects of participants in a séance could create the logical equivalent of a bodiless person through what might be called "group consciousness", investigators who had participated in successful séances decided to invent a "ghost" with whom they would then communicate in their sessions. They thoroughly familiarized themselves with a fictitious character they thought up and found that the raps on the table were correct answers to their questions about the identity of their "ghost" (one rap for yes, two raps for no). If the participants were not cheating, we must assume that these responses were triggered by the mind events of the participants. The experiment, reported by Rosemarie Pilkington, was done in 1973 by members of the Society for Psychical Research of Toronto. It took them many weeks of memorizing and intense efforts during the sessions to "activate" their artificial ghost.(8)

The Physics of Encounter shows that anomalous mind events can influence not only atomic events in another person's brain, but can also influence atomic events in the atmosphere, "hardening" the surfaces of interacting atoms so that they can impact on a solid object like a table and produce sounds. The process is similar to the one that produces UFOs, which radar waves register as solid objects in the atmosphere. The process, which I will describe in Part Two of this book, increases the density of effects in interacting atoms.

The human mind possesses the remarkable power to create observable, coherent events by activating, through morphic resonance, a source of hidden energy. That energy is contained in the "participatory universe" described by Wheeler (7/10). It is a universe that is "looking at itself" and "building itself" in the process. Our universe contains the collective unconscious. When matter events occur, our universe is "looking at itself". When a tree falls in a forest, unobserved by humans, that event, too, is part of the collective unconscious.

The entanglement of mind and matter in our everyday experience is very much in evidence in the world of flowers and other plants. Increasingly, scientists investigating paranormal phenomena are showing an interest in the measurable reactions of plants to the mental events of humans in their vicinity. Measuring devices like those used in lie detectors have proved useful in investigating the processes in the cellular structures of plants. The devices are attached to the plants and measure electrical impulses. These clearly vary, depending on what the plant "experiences". A harmonious environment, like happily playing children, creates a different pattern of impulses in flowers than, for example, the environment in a hospital room for depressed, psychologically disturbed patients. Comparisons between two sets of planted flowers that are both sufficiently watered suggest that people who talk to their plants stimulate their growth.

Even more startling is evidence that plants seem to react to human intentions. As reported by the German science author Gerd Schuster, an investigator named Cleve Baxter wanted to see how a plant would react if he held a burning match under one of its leaves. Before Baxter had lit the match, he noticed that, judging by the suddenly changed electrical impulses, the plant seemed to be "shocked" by his intention. On another occasion, Backster, noticed that after he had just destroyed a petunia for an experiment, a "fellow petunia" reacted violently when he approached it.(9) Can plants read the minds of people?

Thoughts can influence the physical processes in plants in the same way that they influence the processes in another person's brain. In the latter case, the thoughts may be replicated in the other person's mind and the anomalous phenomenon is called *telepathy*. But the effect could also be the result of *psychokinesis*, which is defined as the influence on physical objects by thought processes. As interpreted here, psychokinesis includes the influence of the mind on the motion of elementary particles in the atoms of physical objects. This is the same effect that allows the mind to bend spoons, as described by Pilkington (6/15). The process briefly changes the inner structure of an object. The substance of the spoon is softened and can be deformed. Other examples of psychokinesis are presented in Part Two of this book.

The mind can cause such events when it is *focused*, which enables it to tap into the hidden energy of the collective unconscious. In Roger Nelson's Global Consciousness Project, a consciousness field creates an observable departure from random data when many people become "deeply absorbed in one focus" (8/3). The concept of "focus", in this context, is distinctly dialectical. In Nelson's project, the departure from randomness occurs in every one of the random event generators installed around the world. The effect of the consciousness field is not focused on any specific device. The field is a surge of energy throughout space. To use the term "focus" in this context seems to be a contradiction in terms.

The Physics of Encounter suggests that the focus, in this context, is the specific *quality* of the energy surge. The energy of a consciousness field is the hidden energy of quantum vacuum fluctuations. The energy effect is focused not so much in terms of a specific location in space, but in terms of *structure*. The hidden energy in the wave pattern of an expanding VE-surface "zooms in" on a corresponding wave pattern and the two patterns enmesh in morphic resonance. The resonance occurs where there is structural identity. The process corresponds to what physicists call non-local proximity. The vacuum energy in the expanding VE-surfaces is released at points of encounter. If there is no resonance, there is no encounter.

The wave patterns are information structures in a consciousness field and in the collective unconscious. Where appropriate, I will use the two concepts interchangeably, with the understanding that a consciousness field contains the effects of mind events shared by humans who are alive. This energy field is, by definition, part of the *collective unconscious*. The terms *group consciousness* and *consciousness field* are equivalent, but the former is more appropriately used for smaller groups such as the participants in séances, whereas the latter best describes the effects of mind events in a specific cultural setting or historic era.

The location of any random event in the world of elementary particles can become the focus of a hidden energy surge if that location is a point of encounter between wave patterns interacting in morphic resonance. The wave patterns in the surfaces of VE-spheres, as I explained, consist of layers of QV-bubbles expanding and collapsing as the surfaces expand. The layers correspond to the superpositions in the theory of quantum mechanics. As interpreted here, they become *coherent superpositions* when they are interact in morphic resonance. As explained by Gary Zukav, a coherent superposition "is not simply the superposition of one thing on another. (...It is...) a thing-in-itself which is as distinct from its components as its components are from each other." By using the mathematics of quantum mechanics, Zukav pointed out, "we can calculate the form of this thing-in-itself, this coherent superposition of possibilities (...that exist...) for any given time".(10)

As interpreted by the Physics of Encounter, this means that when the expanding surfaces of VE-spheres interact and create a new *thing-in-itself,* this new reality is a "new mix" of coherent superpositions, a reality that is distinct from what existed before. In other words, what occurs is a change in the *quality* of what is observed or experienced, not just a quantitative change that results from piling additional "things" (effects) on top of one another. This is what happens when a UFO is observed. What was there before, for example a misinterpreted atmospheric disturbance, is changed by the coherent superpositions in the expanding surfaces of VE-spheres. This changes not only the particle interactions in the atoms of the brain but also builds a new event from the misinterpreted event in the atmosphere.

Coherent superposition consists of a multitude of mind events that have something in common. The shared mind events are created in a specific cultural environment. When the exciting possibilities of space travel began to capture the imagination, shared expectations and similar mind-sets increased the probability of misinterpreting normal events in the atmosphere. Lots of people began seeing UFOs.

The UFO phenomenon is created by the hidden energy of the collective unconscious. The phenomenon is not triggered by a willful act of the human mind. It is a subconscious process, like many other anomalous events. They are all energized by the quantum vacuum fluctuations of the collective unconscious. But some anomalous events are associated with a conscious intent. They are the events that occur in séances, for examples, or during demonstrations of psychokinetic prowess (making objects move through the power of the mind), or during "spoon-bending parties" like those described by Rosemarie Pilkington. In those cases, researchers agree, success depends on the ability to attain a state of mind that facilitates the influx of hidden energy. As Pilkington explained (6/15), " it is of primary importance to prevent the conscious mind from interfering (...), for example by occupying it with 'noise and nonsense' (...)." The participants in the spoon-bending party, in her account, "sat around relaxed and in high spirits", they "laughed and joked and chanted (...)."

This corresponds to the kind of dialectical "focus" described above. On the one hand, a willful intent pervaded the party spirit. On the other hand, it was not a grim determination or conscious preoccupation. The mood was playful and relaxed. The participants just waited, with an open mind, to "let things happen".

In experiments demonstrating the possibility of influencing the motion of mechanical target devices, researchers reported that the best results were achieved by participants who tackled the task as an enjoyable game, playfully and relaxed. Robert Jahn, in a collaborative article describing "noteworthy performances in influencing the movements of a little robot on wheels", mentioned that the experimenters had enhanced the "whimsical attractiveness" of the robot by encasing it in a housing resembling a miniature Zamboni machine with a toy frog perched in a driving position.(11)    What counts in such experiments, according to Jahn and Dunne, is a "sense of 'resonance' or 'bond' with the machine, (...) of 'falling in love' with it, of 'having fun' with it."(12)

The physicist William G. Roll pointed out that the ability to tap into the vacuum energy of the universe seems to be comparable to the mind state of Buddhist monks.  "It is sometimes reported that practitioners of yoga and meditation develop psychokinetic abilities. (...) The Buddhist concept of reality as *sunyata*, (...) with which you may unite if your mind is emptied of particulars, is not unlike the idea of a vacuum as an infinite field of energy and consciousness.  (An anonymous Zen Buddhist has called Zen 'the vacuum cleaner of the mind'.)"(13)

A mind that is "emptied of particulars" and avoids "distractions and analytical thought", as Roll and other researchers have put it, increases the probability of bonding, in morphic resonance, with an outside source of hidden energy.  In Recurrent Spontaneous Psychokinesis (RSPK), popularly dubbed Poltergeist events, the strong emotional fixation of a disturbed mind taps into the same source of energy.  The subconscious fixation excludes analytical thought.    It corresponds to the focus in Nelson's Global Consciousness Project.  In psychokinesis, emotional desires and serious intent are not opposites.  Playfulness, as described by Pilkington and Jahn, can complement earnest efforts.  (Remember how earnestly absorbed you were in the games you played as a child.)

The bonding with a field of consciousness that absorbs and empowers the experienced "self" occurs in its most gripping manifestation when the human mind faces the reality of physical death.  The Buddhist concept of the ultimate reality, mentioned by Roll, and all other world religions express this deepest of all human realizations.  Recent advances in medical science have widened the possibilities of investigating a phenomenon that is highly relevant in this context: the so-called *Near-Death-Experience* (NDE).  It occurs, as I've already said, while a human body is clinically dead for a limited period of time. After doctors succeeded in averting final death and the brain started functioning again, many patients reported that their mind seemed very much "alive" while their body was physically dead.

In a documentary aired by a German radio station, Christian Stahlhut mentioned the experiences of an architect who had written two books about his own NDE. A large number of witnesses had confirmed everything he had seen while his mind hovered above his lifeless body. From that location, he said, his mind then traveled deeper into a different kind of reality. In one of the early interviews he had given, according to Stahlhut, the architect, Stefan von Jankovich, groped for words in his description of what he experienced. "Then, in that situation, when I got there, the sun once again flooded me, the light, and then I was somehow, you might say, enlightened (...). Suddenly, I understood a lot of things I did not know before, the whole macrocosm, the meaning of the macrocosm, the meaning of creation (...). Now there is a veil again, the curtains have been drawn, this mystical experience is inside me, always strong and alive, but I cannot express it and no longer understand. In the same way, I understood the entire galaxy, the entire earth as a living being, the whole solar system as a living being. I am a part of the sun, I am a part of the holistic earth principle (...). I understood who I am, I am an energy of God (...) and I understood the whole system of the chakras, I understood the DNA code, back then, which was 12 years before Crick and Watson were awarded the Nobel Prize for what they discovered. I understood Yin and Yang, the endless coming into existence and fading out of existence. At the time, I understood everything and now everything is gone. And we have to struggle upward again."(14)

The concepts that von Jankovich used in the description of his NDE are important in the context of the scenario proposed here. The term *macrocosm* denotes the world of physical objects. The *holistic principle* is the view that an organic or integrated whole cannot be understood simply through an understanding of its parts. A *chakra* is one of the body centers that are considered sources of energy for psychic or spiritual power. *DNA* is a substance that is like the brain of a biological cell, telling it how to grow and how to divide into new cells. DNA gives biological matter the pattern or blueprint for its future development.(15) *Yin and yang* are paired forces in Chinese philosophy. Yin is the passive principle in the universe and yang is the active principle, a source of light and heat.

The sensation of "understanding everything" during a Near-Death-Experience reveals that the boundary between the self and non-self reality is not an absolute boundary. The mind of a clinically dead person experiences itself as a part of everything it sees and understands in a space that is not the space in which human beings live their lives and in which their bodies die.

*Who the living would explain*
*He must enter death's domain.*
Christian Morgenstern

Among the images often seen during an NDE are the images of relatives who had died. As these otherworldly realities communicate, wordlessly, their thoughts and feelings, and what went through their mind when they died, these mind events are experienced as if they had always been one's own. In some of the investigated cases, according to Stahlhut, the person having an NDE had not known, before the NDE occurred, that these relatives were no longer alive. Nobody had informed them that the relatives had died. Even more remarkable is the fact that during an NDE, even blind persons can see, from an out-of-body location, the medical efforts to restore their life. They later correctly described the details of what they saw.**(16)**

We can and we should conduct our everyday lives in keeping with these insights into the deeper reality of our universe. The lessons we can draw from this knowledge have been confirmed in experiments by scientists from various disciplines. The experiments showed that the shared thoughts of large groups (wishes, hopes, feelings) can influence the human activity of many others. The political scientist Courtney Brown has argued, based on his assessment of these experiments, that the influence of mind events on human behavioral systems results from the same type of process through which the mind influences random-number generators. Referring to the creation of particles through acts of observation, he pointed to the "understanding that consciousness (...) can apparently 'force' physical objects to manifest into physical reality despite a prior existence as a less-than-physical probability (...)" and that there is "no broadly accepted reason why this process should break down at the macro level experienced by human society".**(17)**

The physicist John Hagelin followed a similar line of reasoning and flatly stated: "Published research in leading, peer-reviewed scientific journals has shown that collective meditation reduces violent crime, social conflict, and other manifestations of acute social stress." He, too, suggested that some of the fundamental processes currently discussed by theoretical physicists may account for what he calls the "field effects" of consciousness.**(18)**

All of the above facts and arguments show that the common elements of various religious outlooks, meditation as well as prayers, should be taken just as seriously as the scientific research in our day and age. They also show that many elements of the model of mind/matter interaction proposed by the Physics of Encounter are already in place in theoretical physics.

In summary, the model I am proposing suggests that the *vacuum energy* in the surface of expanding VE-spheres provides the energy from an "outside source" demanded by the laws of physics for the creation of structure, or information, where none existed before. The location of that source of energy is the *quantum vacuum* in the QV-bubbles that accumulate in layers (superpositions) in the surfaces of VE-spheres as these surfaces expand. The superpositions represent what Courtney Brown called a "less-than-physical probability". The probability becomes physical reality when the surfaces of VE-spheres encounter one another, in morphic resonance, within a coupled map lattice. The expanding surfaces of VE-spheres are probability waves, or quantum mechanical wave functions. The points of encounter between VE-surfaces are the locations where, as Michael Manthey put it (3/3), we "get something from nothing".

The metaphorical images proposed here account for the effects of consciousness by showing how oscillating strings create the basic ingredients of consciousness (qualia) in the quantum vacuum of QV-bubbles. The strings are conceptual constructs introduced by a number of theoretical physicists (Hagelin is one of them) to unite the two monumental theories of physics, relativity theory and quantum mechanics. The Physics of Encounter enlarges the common ground of these theories and enables them to shed more light on the nature of elementary reality.

A fresh look at the processes that occur inside atoms and during the interactions between them is necessary. The second part of this book describes the specifics.

*NOTES*
(1) Lemonick (2005), p. 53
(2) Nelson (2002), p. 556
(3) Ibid, pp. 553 and 566
(4) Vallee (1969)
(5) Bonvin (2005), p. 132
(6) Ibid, p. 79

114

(7)  Ibid, pp. 391-413
(8)  Pilkington (2006), pp. 211-213
(9)  Schuster (1989), p. 262
(10) Zukav (1979), p. 285
(11) Jahn *et al* (2007), pp. 27-46
(12) Jahn and Dunne (1997), p. 210
(13) Roll (2003), pp. 81 and 83
(14) Stahlhut (1993), 28
(15) *Popular Science* (1968), p. 383
(16) Stahlhut (1993), p. 29
(17) Brown (2007), p. 8 (Abstracts)
(18) Hagelin (2007), p. 15 (Abstracts)

# PART TWO

## Anomalous events - categorized processes

## *Chapter 9*

**The powerful light of UFOs. Photons and wave functions. Anomalous plasma and atoms of oscillating size. The trance during the observation of a UFO. Photographs of the "unidentified flying objects".**

Some anomalous events are easily identified because of the abnormally large amount of energy released when they occur. There is plenty of evidence, for example, that UFOs are "real", in a strictly physical sense. The most compelling evidence is the light they radiate. In many cases, the power output of this light is enormous. It can be stronger than the energy that is unleashed in a bolt of lightning. Amazingly, the powerful light of a UFO is present over a longer period of time, and not over in flash, as is the case with lightning. Physicists have no satisfactory explanation for this anomalous energy.

The explanation provided by the Physics of Encounter will be presented in this chapter. First, some undisputed facts. The physicist Jacques Vallee described an incident that took place in France in 1978 and was thoroughly investigated by GEPAN, the French government's official UFO investigation task force (Groupe d'Etude des Phénomènes Aériens Non-identifiés). The incident took place near the town of Arcachon, where a UFO, observed by several witnesses, appeared in the middle of the night. Its brightness triggered the photocells that control the lights of the whole town. The photocells automatically switched off all street lights, as they do when daylight arrives. Scientists later dispatched by GEPAN made measurements in the field to establish that all witnesses had observed the same object. They interviewed the witnesses and had them point a theodolite to the places where the UFO had appeared and disappeared. A theodolite is a surveying instrument used to establish precise locations by measuring vertical and horizontal angles. The witnesses were given a set of color samples from which they made a selection corresponding to the phenomenon they had observed. They described the object as oval and red, or as a large orange ball. The UFO, estimated to have been between 135 and 480 meters away from the photocells that controlled the street lights, exposed them to a luminosity that exceeded their threshold ($10 \text{ mW/m}^2$). Vallee calculated that the luminosity was between *2.3 and 29 kW*, depending on the distance of the UFO.**(1)**

Vallee published several documented incidents with multiple witnesses extracted from a larger sample where luminosity or power output data could be obtained. One case, in 1965, involved two French submarines, the Junon and the Daphné, over which "a large luminous object arrived slowly and silently from the west, flew to the south, made three complete loops in the sky over the French vessels, and vanished like a rapidly extinguished light bulb (...). "

According to Vallee, the person who reported the case to him, Michel Figuet, was the first *timonier* (helmsman) of the French fleet of the Mediterranean. "He observed the arrival of the object (...) (and) had time to go up to the conning tower, where he took six pairs of binoculars and distributed them to his companions. There were three hundred witnesses, including four officers on the Junon, three officers on the Daphné, a dozen French sailors, and personnel of the weather observatory."

"All witnesses aboard the Junon saw the object as a large ball of light or a disk on edge arriving from the west at 9:15 p.m. It was the color of a fluorescent tube, about the same luminosity as the full moon. (...) It moved slowly, horizontally, at a distance estimated at ten kilometers south of the ships (...). It left a whitish trace similar to the glow of a television screen. When it was directly south of the ships the object dropped toward the earth, made two complete loops, then hovered in the midst of a faint 'halo'."

"Mr. Figuet told the author that he had observed the last part of this trajectory through binoculars; he was able to see two red spots under the disk. Shortly thereafter, the object vanished in the center of its glow 'like a bulb turned off'. (...) (Later) the halo reappeared at the same place, and the object seemed to emerge as if switched on. It rose, made two more loops and flew away (...). The next day Mr. Figuet compared notes with a communications engineer who had observed the same object from the Navy fort. Together, they called the weather observatory at Fort-de-France. The man who answered the phone had also observed the object." Since there was agreement among the observers that the object had approximately the same brightness of the full moon and was situated about 10 kilometers away, Vallee computed the total luminosity of the object as *2.3 megawatts*.**(2)**

To obtain luminosity data on UFOs, Bruce Maccabee analyzed a photograph taken in 1956 about 20 minutes before sunset by a Canadian Air

Force Pilot of a very bright, disc-like object that was remaining stationary in the midst of a mass of cumulous clouds called thunderhead. The analysis suggested that the object radiated *more than a gigawatt of power* (!) within the spectral range of the film. The object was well over 100 meters in diameter. Maccabee argued that the object could not be explained as *luminous plasma*, a high-temperature gas containing electrically charged atoms. Its calculated radiation levels, as reported by Maccabee, "make it difficult to sustain the argument that the object was plasma as ordinarily understood. Neither the origin of the plasma, its size, its duration nor its total power output seems consistent with known plasma phenomenology. Its location near a huge thunderhead suggests that some transient electrical effect of the storm (e.g. lightning bolt) might have created it, but this would not explain its duration of many tens of seconds. Assuming a minimum estimated sighting duration of 45 s and a minimum estimated distance of 6 km the energy radiated would have been (...) about two orders of magnitude greater than the energy in a typical lightning stroke. (...) An energy density this great (...) implies that there must have been a continuous power source either within our outside the plasma."

"An alternative phenomenon which may be considered is ball lightning or 'kugelblitz'. Ball lightning has some characteristics of an ordinary plasma but it exists for much longer periods of time and therefore must have some (unknown) storage mechanism 'built in' or else it is continually powered in some (unknown) way by an external power source. (...) Unfortunately, there is no good theory that explains how a typical ball lightning could exist."(3)

The key phrase in Maccabee´s reasoning is that a "continuous power source" must exist "either within or outside" such luminous objects. As interpreted here, that power source originates in the quantum vacuum fluctuations of space, which contain the collective unconscious. As demanded by the second law of thermodynamics, an outside source of energy is necessary to create structure, or coherence (in short: a new physical reality) where this did not exist before. The Physics of Encounter suggests that this power source is the hidden energy of the collective unconscious. It triggers the observation of anomalous events, by one person, or several, and these events can then be observed, and are often enhanced, by the mind events of many other persons. Since quantum vacuum fluctuations occur, in Puthoff's words, "throughout space, even in so-called empty space", the anomalous power source exists both inside and outside the luminous object.

Here's how the proposed model of mind/matter interaction explains the abnormally large power output of UFOs (and ball lightning):

Since quantum vacuum fluctuations create mind events as well as matter events, what happens in the mind also influences events outside the mind. Everyone has an "extended mind". In the case of UFOs, anomalous mind events influence the size of atoms outside the brain. Like all mind events, anomalous mind events are created by interacting VE-spheres. As the expanding spherical surfaces interact, the QV-bubbles in these surfaces expand and collapse at points of encounter. This causes the fluctuations of the vacuum energy in the QV-bubbles, which become increasingly numerous within the expanding surface of a VE-sphere. The vacuum energy in the surface increases as the surface expands.

Anomalous mind events create abnormally large VE-spheres. Let's neglect, for the moment, the psychological aspects of the process that makes this happen and look at the physics of the process. All interactions in the world of particles result from encounters between the surfaces of VE-spheres, or probability waves. This includes the interacting elementary particles that, together, are the substance of atoms.

The abnormally large amounts of energy associated with UFOs and ball lightning result from abnormally large fluctuations of vacuum energy. Large amounts of vacuum energy are contained in large VE-spheres because their surfaces contain a large number of QV-bubbles. When such VE-spheres expand and collapse, due to encounters in morphic resonance, the size of the atoms affected by anomalous mind events oscillates accordingly. From one instant to the next, the atoms are either abnormally small or abnormally large. When atoms are abnormally large, they are like the so-called Rydberg atoms. As reported by the science author Alexander Stirn, the physicist John Gilman proposed a theory to explain the power output of ball lightning. It involves atoms that are abnormally large, but do not fluctuate in size.

In Gilman's theory, the orbital electrons in abnormally large atoms are unusually far away from the atomic nucleus. According to Stirn, the distance can be up to several centimeters, which is huge considering that normal atoms are almost invisibly small. The space inside these anomalous atoms is filled with enormously intense activity. The atoms are, to use the technical term, extremely "agitated". The theory was published in 2003 in the prestigious professional journal *Applied Physics Letters*, so it wasn't just some wayward, speculative idea. Nevertheless, it was immediately challenged because, so the counter-argument, the heat energy of the agitation would destroy the individual atoms, which would then no longer be separate physical entities. As Stirn explained, you would wind up with what you had to begin with: an extremely hot, inexplicable kind of plasma.**(4)**

The Physics of Encounter avoids this dilemma by suggesting that the atoms in an anomalous plasma oscillate between being abnormally large and abnormally small. The oscillation occurs in the space between the interconnected atoms, where magnetic forces and the exchange of electric charges energize the process that creates the bond between the atoms. The process is similar to the exchange of photons between atoms at separate locations in space. That exchange is an electromagnetic wave. In an anomalous plasma, just as in normal substances composed of bonded atoms, the respective electron orbits of the atoms are "intertwined". Electric charges are transferred between the interacting atoms.

In the scenario proposed here, the space between the interconnected anomalous atoms is filled with an abnormally large amount of vacuum energy. It is the same vacuum energy that exists inside an electron shell and triggers the quantum jumps of electrons from one orbit to another. They are jumps from one *energy state* to another. The energy state of an electron inside an atom depends on the distance between the negatively charged electron and a positively charged proton in the atomic nucleus. Opposite electric charges attract each other. Electrons jump, or "fall", into a lower orbit, or energy state, when they are attracted toward the nucleus. They jump, or "rise", into a higher energy state when they increase their distance from the nucleus. These jumps between energy states, as interpreted here, are triggered by fluctuations of vacuum energy that influence the electric charge of particles in the atomic nucleus. Theoretical physicists have established that the protons in atomic nuclei oscillate. For an instant, they lose their positive electric charge and become neutrons (articles without an electric charge). Then they immediately become protons again, in constant alternation.

The proposed scenario is supported by Ervin Laszlo's argument that, under certain conditions, vacuum energy acts on electrons orbiting atomic nuclei. The energy of the photon emitted when an electron falls into a lower "orbit", or energy state, corresponds to the energy that the electron loses during such a quantum jump. An anomalous energy change caused by vacuum energy, Laszlo argued, occurs when photons exhibit a so-called *Lamb shift*. It is a sudden change in the energy of light.**(5)**

In a Lamb shift, the energy change of the photon emitted from an atom is minimal. If atoms are abnormally large, however, the light waves created by the emission photons have a much higher energy. The energy lost by an electron during the jump or "fall" into a lower energy state is larger than normal because the event is an abnormally large quantum jump. The energy of an emitted photon is correspondingly large. A destruction of the individual atoms due to the heat of the high-energy oscillations of the light emitted from the atoms would not occur because the atoms instantly contract into an abnormally small size, and then expand again, in constant alternation.

As interpreted here, a photon, at the instant of its creation, is a point of encounter between the expanding surfaces of VE-spheres. In other words, it is a point-particle that has absorbed vacuum energy and becomes a QV-bubble, which then expands into a new VE-sphere, as explained in chapter three. As soon as the photon is created and emitted from an atom, it is no longer a particle, but part of a light wave. It is part of the physical system that consists of interacting VE-spheres, expanding and collapsing at points of encounter. The particle effect of a photon becomes apparent again at the instant the photon is absorbed into an atom. It is a pinpointed energy effect. When the event occurs, the photon ceases to exist. Its energy is absorbed into the atom.

The interaction between abnormally *large* atoms produces the strong *power output* of the light emitted by UFOs and ball lightning. The anomalous effect associated with the interaction of abnormally *small* atoms is abnormally strong *radiated* light. Both effects are created by atoms that are no longer normally bonded. The bonding substance is an anomalous *plasma*.

What is the difference between the power output of a light source and the strength of the radiation? The radiated energy of a light wave is large if a large number of photons is emitted per unit of time. In the scenario proposed here, the radiation is abnormally strong when abnormally small atoms interact. The high number of encounters between the VE-surfaces in the respective electron shells of the interacting atoms corresponds to a high frequency of photon emission. Physicists express the energy of a light wave in terms of the frequency of the oscillations in electromagnetic radiation. A high frequency corresponds to a small wavelength of the radiation. Frequency and wavelength, as interpreted by the Physics of Encounter, are determined by the expansion and collapse of VE-spheres encountering one another in morphic resonance. The wavelength of the radiation is the radius of the interacting VE-spheres.

High-energy light waves are like a spray of photon-bullets fired in rapid succession into all directions of space. The photons are of exceedingly short wavelength. Below a certain wavelength, however, the waves are no longer visible as light. The waves that are *shorter* than the visible wavelength of light are invisible ultraviolet radiation, as in X-rays. On the opposite end of the scale, infrared radiation is invisible because it has a wavelength that is *longer* than the visible wavelength of light.

The power output of light is not the same as the energy associated with the wavelength of electromagnetic radiation. The power output is measured in watts. It is the power of the luminance measured inside an illuminated area over a period of time. A reference system for comparing the strength of a luminance could be the number of watts per square centimeter, for example. To illuminate a large area requires more power, obviously, than to illuminate a relatively small area. As described at the beginning of this chapter, a UFO that appeared over a French town in the middle of the night was so bright that all of the street lights of the town were automatically shut off. The UFO had illuminated a large area of the town with light that was as bright as daylight.

An additional yardstick for determining the power output of a source of light is the length of time during which the light continues to shine with undiminished brightness. The engine of a rocket taking off into space creates a bright stream of light that trails the rocket for a certain time until the fuel is used up, or until the engine is shut off when the rocket has reached the required altitude. The amazing aspect of UFOs is that some of them shine so brightly over such a long period of time that physicists are at a loss to explain where the UFOs get the "fuel" to keep shining with such steady power.

The fuel is the plasma that interconnects anomalous atoms. The atoms oscillate between being abnormally small and abnormally large. The anomalous plasma is continuously re-created by the hidden energy of anomalous mind events that occur as long as UFOs (or ball lightning) are observed.

When we observe an object in space, photons are emitted from that object and their effects are transmitted to the atoms of our brain. Because of the wave/particle duality of photons, even *normal* light waves cannot be adequately explained by mainstream physicists. What is the reality of a photon when it is a point-particle? What is the reality of the photon before it impacts on an atom? How do dimensionless points influence the world of matter? The data gathered in many years of research have given physicists no choice other than to accept the vexing wave/particle duality. Things have not changed much since the days of Louis-Victor de Broglie, who received the Nobel prize in 1929 for his theory of matter waves. According to the physicist Ernst Zimmer, de Broglie exclaimed: "We would know so much more if we only knew what rays of light really are!"(6)

In the proposed scenario, a photon is created from the vacuum energy in the nucleus of an atom and is emitted from there. It expands into space as the surface of a VE-sphere. The photon does not exist as a particle in space until it creates a new instant in time when it is absorbed into a different atom. Before that instant, it exists as vacuum energy in the invisible surface of a wave. It is what physicists call *potential energy*. As part of a wave, a photon is an expanding spherical surface that transmits vacuum energy from one atom to another.

My metaphorical comparison of photons with expanding spherical surfaces that are part of a wave is based on similar images used by physicists. In their terminology, the invisible surfaces that expand like waves are *wave functions*. That term refers to the mathematics that allow physicists to calculate the probability of an event. In this case, it is the probability that an event caused by a photon will occur at a certain location. The physicist John Cramer put it this way: "A photon emitted from a source has no directional preference. The emission creates a spherical wave function that expands like an inflating bubble."(7)

The "bubble", as interpreted here, is a VE-sphere that has expanded from a QV-bubble. The expansion of the VE-sphere is energized by the steadily increasing number of QV-bubbles that exist within the surface of the VE-sphere. Each QV-bubble contains, for one instant only, the qualia of the observer's consciousness. The conclusion that photons are VE-spheres is supported by the remarkable fact, mentioned by Stahlhut ((8/16), that during a Near-Death-Experience (NDE), the mind of a blind person can observe, from an out-of-body location, the activities of doctors as they successfully stave off final death.

As a VE-sphere, a photon can allow a blind person to see, during the extraordinary moments of an NDE, because a brain contains bio-photons, as described by VanWijk (4/2) and other biophysicists. When photons are absorbed into atoms, some of the vacuum energy in the surfaces of

VE-spheres accumulates in the nucleus of the atom. There, this energy becomes what I have called the "inner light" of consciousness inside an atomic nucleus. This light is the energy of particles called gluons. In terms of energy interactions, physicists have called gluons the inner-atomic equivalent of photons. The exchange of photons between atoms establishes a link between these atoms. In an atomic nucleus, the exchange of gluons establishes a link between the events inside the nucleus.

When an atom absorbs a photon, that is not an isolated event because what we see, thanks to photons, originates in interacting atoms that emit these photons. Our consciousness straddles countless atoms emitting and absorbing the photons of a light wave.

Two aspects of this mind/matter scenario are important here. As I explained in chapter two, atoms are mostly *empty space*. They are complex configurations of VE-surfaces interacting at points of encounter. Clusters of the QV-bubbles in these VE-surfaces, expanding and collapsing, are the dynamic particles inside the atoms. Secondly, photons are not like other QV-bubbles expanding into VE-spheres. The VE-surface of a photon distinguishes a photon from matter particles, which "stay put" within a more or less clearly defined location in space. The characteristics of a VE-surface are determined by other VE-surfaces that interact with it and influence its effects. The specifics of the processes that distinguish the many different kinds of particles from each other, and from the point-particles of light, are described in Part Three of this book (chapter twenty-one). They are not important for a basic understanding of the process that creates the anomalous power output of UFOs.

The process, as I said, involves anomalous mind events that create atoms of fluctuating size. The focal point of the events that bring forth consciousness is, of course, the brain. The atoms of the brain, and the bio-photons exchanged between them, are primarily involved in the continuity of the mind events that characterize the identity of a human being. They are part of what the physiologist John Eccles called the "neural machinery" of the brain. The photons processed by this "machinery" are VE-spheres of normal size. Anomalous matter events such the sudden appearance of a UFO are caused by abnormally large VE-spheres originating *inside* the brain and encountering one another *outside* the brain.

The creation of abnormally large VE-spheres through anomalous mind events occurs because, as Eccles argued (7/3 and 7/4), the mind is an "independent reality" that acts "in accord with its attention and its interests (...)". The mind, Eccles said, "is actively engaged in searching and probing through specially selected zones of the neural machinery (...)".

As interpreted here, this means that the independent mind deals in its own way with an event it observes in the atmosphere. The observation causes a quantum mechanical event in the brain, and the mind then engages in searching and probing through that zone of the neural machinery. If the mind misinterprets the event in the atmosphere and sees it as something that might be a UFO, the emotion and the intellectual excitement creates an energy "focus", as described by Nelson (8/3).

That focus shuts out, for the moment, the observation of other events. The observer becomes preoccupied with the emotional experience. Expressed in terms of the qualia of consciousness created in the QV-bubbles of expanding VE-surfaces, this means that the spectrum of experienced qualia is narrowed. The emotion predominates the mind. The number of encounters between the surfaces of VE-spheres that produce the wide variety of qualia is reduced. Since VE-surfaces expand until they are encountered, the reduced number of encounters allows the VE-spheres to expand into a larger size. The focus on the misinterpreted event and the temporary exclusion of other aspects of reality can create the trance-like state of mind, or self-hypnosis, that is often associated with the observation of a UFO.

The surfaces of abnormally large VE-spheres consist of the *coherent superpositions* described by Zukav (8/10). They are "a thing-in-itself (...)", as he said, and "we can calculate the form of this thing-in-itself, this coherent superposition of possibilities (...) for any given time". As interpreted by the Physics of Encounter, the form of this thing-in-itself can be the form of a UFO. It may look like a spaceship to the observer, but it does not have any distinct form in the space of measurable physical events. It is, to put it bluntly, a shapeless blob of quantum mechanical interactions in the atmosphere, an anomalous source of light.

The highly compacted mass of what appears to be a UFO is often detected on radar screens. Paradoxically, that is not always the case. The reasons for this have to do with the distance between the radar screen and the UFO. For similar reasons, attempts to take photographs of such "unidentified flying objects" are sometimes successful, but quite often they are not. Even when the light from what appeared to be a distinct form was clearly visible in the viewfinder of the camera, no such image was imprinted on the film. That's because the image is a *coherent superposition*. It exists within the expanding surface of VE-sphere and is, as Zukav said (8/10), a "thing-in-itself", a possibility at a certain point in time.

The effect of the image is like the effect of a photon. It exists as a possibility, as the potential energy in a quantum vacuum. Coherent superpositions are the layers of QV-bubbles in the surface of a VE-sphere. To capture the "thing-in-itself" on film, therefore, the shutter of the camera would have to be opened at a precise instant in time and at a certain distance from the UFO.

When anomalous events occur, the amount of energy involved in an act of observation is amplified by the hidden energy of the collective unconscious. This is what creates abnormally large VE-spheres. When the surface of such a VE-sphere is encountered in morphic resonance, this releases a large amount of the hidden vacuum energy that is contained in the surface. A new VE-sphere, again abnormally large, is created at the point of encounter.

Because the VE-sphere is so large, the location of the UFO created by the hidden energy may be quite far away from the location of the observer. The coherent superpositions in the surface of the VE-sphere will have an effect on observers (and their cameras) only if these are located at the *boundary* of the abnormally large VE-sphere, because this is where the encounters occur. The observed form of the UFO exists as a temporarily stable effect only in that surface, not in the neighboring regions of space. Through acts of observation, the surface of the VE-sphere is continuously re-created at that location for a certain period of time.

The surface of the VE-sphere is an invisible reality, like the surface of a photon *before* it impacts on an atom. When the impact occurs, a pinpointed effect becomes an observable physical reality. The effect is pinpointed in space and in time, but *where* that happens, at a point in the spherical VE-surface, and *when* that happens, is unpredictable. VE-spheres expand in the dimensions of space and in the dimension of time. Quantum mechanical events are random events. Since the observed form of the mysterious object consists of patterns of VE-surfaces that contain the same potential energy as those in the wave patterns of light, UFOs can sometimes be photographed. But the form that is photographed exists in the "other reality" of our universe. It is the hidden energy in the "empty" space of quantum vacuum fluctuations.

126

*The relationship between form and emptiness*
*cannot be conceived as a state of mutually exclusive opposites,*
*but only as two aspects of the same reality,*
*which coexist and are in continual cooperation.*
Lama Govinda

*NOTES*
(1) Vallee (1998), pp. 354-356
(2) Ibid, pp. 348-350
(3) Maccabee (1999), pp. 205-206
(4) Stirn (2003), p. 12
(5) Laszlo (1996), p. 4
(6) Zimmer (1964), p. 99
(7) Cramer (2006), *conference presentation*

# *Chapter 10*

**"Materialization". UFOs and the phantom images created in a séance. Mirror neurons and the jolts in a railroad car. The "embarrassing infinities" in a Higgs field.**

When UFOs are detected by radar, they produce blips on radar screens that are suddenly there, as if a massive object had suddenly materialized. Equally puzzling is the fact that a UFO may suddenly vanish, light a light bulb is turned off, as Vallee put it (9/2). The mass of the "unidentified flying object" just seems to disappear.

Mass is a characteristic of matter that gives a physical object its weight. But what seems so simple to grasp, intuitively, is the result of a process at the elementary level of reality that physicists do not yet fully understand. How do elementary particles acquire their mass? When the universe was created, according to current wisdom, it exploded into pure energy and the particles, when they began to form, had no mass. They were like a spray of energy droplets.

That is why the results of the experiment at the CERN laboratory, as I mentioned in chapter four, were so eagerly awaited. The experiment was designed to reproduce the conditions that presumably existed a split second after our universe exploded into existence in the proverbial "big bang". Under those conditions, according to the so-called standard model of physical reality, an elusive particle called the Higgs boson should be detectable by the elaborate measuring instruments at CERN. Physicists had theorized that the *Higgs field*, in which the boson exists, would account for the process through which all particles acquired their mass.

Before I get to the details of this theory, here is an additional example of how an invisible force seems to produce the effects of mass. This happens during *séances*, when a handful of people gather in a darkened room and establish what I have called "group consciousness". Empathy among the participants is needed to get into the swing of things. One of the goals of the group is to conjure up phantom images. As I already explained, the Physics of Encounter suggests that such groups may succeed in tapping into the hidden energy of the collective unconscious.

In a joint effort characterized by a relaxed and playful spirit the participants create for themselves a large variety of variously shaped lights and luminous objects. Quite often hands or limbs of various sizes and colors seem to materialize and then fade away again. As described by Rosemarie Pilkington, the experienced events include "touches, pinches, pulling on clothing, and other tactile phenoma".(1)

The touching, pinching, and related sensations in séances have no recognizable cause. The effects seem to come out of nowhere. Since the effects in séances are often produced by tricksters, parapsychologists have conducted many investigations. Pilkington described one that took place in 1894. The English parapsychologists F.W.H. Myers and Oliver Lodge, and a Polish investigator, Julian Ochorowicz, participated in a séance by the well-known "medium" Palladino. Myers, according to Pilkington, "had been to hundreds of séances and knew every trick in the book (...)". During the séance he investigated, "Myers was 'repeatedly and vigorously' shoved from behind. Lodge changed places with Ochorowicz to see what was pushing him but saw nothing visible, although Myers kept feeling the shoves and complained that 'things which pushed like that must be visible (...)'."(2) According to Pilkington, Lodge was also a physicist and described the effects as "being caused by 'vitality at a distance'. He said he had been touched and had seen others being touched by a 'prolongation or formation' that he experienced 'as if the connecting link, if any, were invisible and intangible, or as if a portion of vital or directing energy had been detached, and were producing distant movements without any apparent connexion with the medium (...)'."(3)

In Pilkington's description of various séances, a particularly talented medium was readily able to produce images as desired by others. The phantom images of deceased persons "would gradually develop from a kind of haze, slowly becoming more distinct, and acquiring clearer individual features, such as facial lines or mustaches, and various textures, such as the silk or other fabrics of their clothing, metal buckles, or leather gloves. (...) The apparitions could change in what seems to be a phenomenon similar to today's computer-generated metamorphosis of faces. (...) In a few instances a face would become more masculine, then more feminine, as if several faces were superimposed over one another. (...) The apparitions (...) would interact with the sitters and generally do what was requested of them, although they would at times display (...) independent behavior."(4)

As interpreted here, the appearance and the motion of the phantom objects and the behavior of the personalized apparitions correspond to the conscious and subconscious desires of the participants in the séance. The phantom image of a person, therefore, "does what it is told". The "independent behavior" is, of course, also the product of desires and expectations, but they are subconscious expectations not recognized as such by the participants. The phenomena observed during séances mirror the mind events of the participants, but the participants do not realize that there is a "connection".

A partial explanation of these experiences is provided by the theory of *mirror neurons*. The theory, developed by psychiatrists, describes the effects of nerve cells (neurons) in the brain. These neurons trigger experienced events that *feel* like actual movements, but in reality they are just mirror images of observed movements. Because the discovery of this effect was quite a sensation, newspapers also carried the story. An article in the *International Herald Tribune* by Carey Goldberg began with questions: "Do you ever feel a twitch in your arm as you watch a baseball player wallop a ball? (...) Do you tense as a TV surgeon slices into an incision? Those are 'mirror neurons' at work. (...) Italian neuroscientists studying monkeys were amazed to discover that (...) the same neurons become active when a monkey actually makes a movement and when it is only watching another monkey, or even a human, make that same movement. It is as if the monkey is imitating - or mirroring - the other's movement in its mind. (...) Mirror neurons are (...) brain cells that control movements. But they show signs of activity not only when a person moves, but also when a person only observes someone else making that movement." The article mentions recently published research "suggesting that the mirror neurons respond not merely to another person's action but also to the intention behind that action".(5)

This corresponds to the anomalous effects described by Schwartz and Russek (5/10), namely that when someone just *imagines* to be staring at someone else, the other person may feel the effect of that mind event. The Physics of Encounter suggests that imagined events can influence not only the brain events of another person by influencing atoms in that brain, but can also influence the atoms where a UFO is observed, or the atoms where a phantom image is observed during a séance. By influencing these atoms, physical events that affect the mind actually occur at these locations. In a séance, they create the feeling of being shoved or touched by the phantom image, or by an invisible force imagined to be at that location.

The feeling of being touched by a "prolongation or formation", as the above-mentioned participant in a séance described it, is like the feeling of being shoved by a massive object. Similar sensations have been described by participants who reported that they saw stones falling from the ceiling and actually felt the impact. The stones seemed to be massive, but they where phantom images. Similar effects are experienced in séances that create images of disembodied limbs. For a few instants, they seem to possess "real" mass. Participants who said they touched the phantom image of a hand, for example, reported that it felt soft and rubbery and then dissolved.

When the mind (the "self") experiences physical reality (the "non-self"), and thereby experiences its own existence, it experiences movement. It may feel pushed or shoved by an invisible reality, as in a séance. The force that "pushes" is the force through which the mind becomes aware of its own existence. Just *seeing* something would not do the trick. By seeing something, the mind may become aware of the *thing*, but not of *itself*. What is needed is an instantaneous connection between one kind of inner event (seeing) with a totally different kind of event (feeling pushed). When that connection is established, the mind becomes what Eccles called a "self-conscious mind" (7/3). A person endowed with consciousness feels: "Hey, something is *happening* to me, and it's not me who's doing it!"

Pushing will do the trick, or pulling. A good example of experiencing the effect of motion, of feeling "shoved", is what you feel when you are sitting in a train that has stopped at a station and then starts to move again. You feel a jolt when the movement begins. Once the train has stopped accelerating and is traveling straight forward at a constant speed, you no longer feel any effects of the motion (neglecting vibrations and the bumps of not totally leveled tracks). That kind of idealized smooth motion is approximated in European trains traveling on state-of-the-art tracks , but -alas- not by Amtrak.

Now imagine that you are in a train traveling in idealized smooth motion, and that the railroad car in which you are sitting has no windows, and that no lights are turned on. You would be completely "in the dark" about the motion of the train. You would not know whether the train is standing at a station or speeding along. Suddenly, you feel a jolt. You would want to know what happened. Did the train suddenly start moving, or did it slow down abruptly? In the situation described here, you could not tell the difference. If you are sure that you were looking toward the front of the train when you sat down, and if the upper part of your body snaps forward when you experience the jolt, you can deduce that the engineer applied the brakes. But if you do not know into which direction the train is traveling, the jolt could also mean that the train abruptly increased its speed and that you are facing the rear of the train.

The interesting thing about this kind of jolt is that you would experience it even if your body, or any part of it, did not move at all, relative to the railroad car in which you are sitting. This means that the jolt is very much like the experience of being pushed or pulled by an invisible force during a séance. I already mentioned the experience of motion that can be created by the force of magnetism. If a magnetic field is applied to your brain, and you keep your eyes closed, you will feel as if you are being "floated" upward and away. During the hypnotic fixation of the mind on a UFO experience, an abnormally strong magnetic force contributes to the experience of being abducted into an alien spacecraft.

A sudden change in motion influences the mass of an object. If a particle moving through empty space is given a boost that increases its velocity, the particle acquires additional mass. So what kind of "substance" does the particle absorb when it becomes "more massive"? The reality of mass and matter is much more puzzling than it seems. To show that the effect of gravity can be created by motion, my high school textbook on physics asked us to imagine an elevator in an endless elevator shaft somewhere in outer space. There is no gravity in outer space, so a person in the elevator would be weightless. But if the elevator accelerates "upwards", and you are in it, you would experience the equivalent of gravity. Your body would be pulled toward the floor of the elevator as if your body weighs something. Weight is the result of gravitational "attraction". Your "weight", inside that elevator, would exist as long as the elevator continues to increase its speed. But if the velocity levels off and then remains unchanged, you will be weightless again and float inside the elevator.

The physical effect of changes in motion corresponds to the *experienced* effects created by *mirror neurons*. They create an effect like the jolt you experience in a railroad car. The instinctive awareness that gravity and motion are involved in mind events is evident in expressions that have developed in various languages. For an idea that "comes to mind", the German word is *Einfall*. It implies that something has "fallen" into the mind. The English word "emotion" also belongs into this category. One other example I will give here is the expression for an experience that "touches us". It is a "moving experience".

A mind event occurs when the "motion" of an expanding VE-surface ends abruptly at a point of encounter with another VE-surface. This creates a new QV-bubble that contains new qualia of consciousness. The moment of encounter is a dialectical reality. Logically, it occurs when the distance between a moving entity and the location where it encounters a different entity shortens and finally becomes a point. Logically, that event cannot happen.

Zeno of Elea, a philosopher in ancient Greece, demonstrated this pitfall of conventional logic with his example of an arrow flying toward its point of impact. You can look at the remaining distance and, in your mind, cut it in half. When the arrow has arrived at that position, you can again cut the remaining distance in half. In your mind, you can continue that process endlessly. Logically, there is always a remaining distance that you can cut in half. But by this reasoning, the arrow would never arrive. Mathematically, the movement toward a point of arrival, as explained in my encyclopedia, is "the sum of an infinite series of geometrically decreasing distances. The series has a finite sum even if it has an infinite number of terms."

What you experience as a jolt, or a mind event, is the effect of something that has "arrived". In French, by the way, one of the meanings of the word *arriver* is "to occur, to happen". In the Physics of Encounter, what arrives at a point of encounter is the vacuum energy in an expanding VE-surface. It is not a question of *what* has arrived, of "this" or "that", as Zukav put it in his explanation of superpositions (8/10). The sum total can be anything and everything. The arrival itself is an event, a zero-dimensional point of impact.

That point corresponds to the endpoint of an oscillating string. All of space is filled with these "strings". They provide the energy of the quantum vacuum fluctuations that are the dynamic element in all of space. The energy exists in the vacuum of "empty" space between the galaxies of our universe and in the tiny quantum vacuum of a QV-bubble, where the oscillating strings create the qualia of our consciousness.

Now let's look at the encounter that creates mass. The encounter is a complex event because the VE-surfaces expanding toward a point of encounter contain the layers of QV-bubbles that are the superpositions mentioned above. One layer is superposed on the other. The smallest bubbles are located in the outermost layer. As QV-bubbles expand, they become VE-spheres and new QV-bubbles are created in the surfaces of these spheres. I described the process in chapter three.

This means that an encounter between VE-surfaces involves a *cascade of encounters* between the QV-bubbles in the surfaces of increasingly larger VE-spheres. I already mentioned that I am using the term "VE-surface" interchangeably with the term "surface of a VE-sphere". The Physics of

Encounter suggests that the cascade of encounters associated with an encounter between expanding VE-surfaces, or probability waves, corresponds to the *Higgs field* that creates the mass of elementary particles. The Higgs field, in this scenario, is the space between expanding VE-surfaces that are about to encounter each other. The final event involving the two surfaces before they collapse, due to the encounter, is preceded by a cascade of interrelated encounters that create particles. They correspond to what physicists call *messenger particles*. In the proposed scenario, they signal the imminent occurrence of what would logically be the more comprehensive final encounter. The final encounter, in this scenario, creates the particle that has been dubbed the "Higgs boson".

The CERN experiment was designed to prove the existence of this particle, thereby confirming the theories of mainstream physicists who have argued, as the Nobel prize physicist Leon Lederman put it, that mass "is not an intrinsic property of particles" but is acquired in a Higgs field through "system configuration".(6)   Lederman's reasoning supports the model of mind/matter proposed here. The system configuration corresponds to the scenario of interacting VE-surfaces, in which the mind of the observer contributes hidden energy to the interaction.

A by-product of the creation of particle mass in a Higgs field through a cascade of encounters between the QV-bubbles in the converging surfaces of VE-spheres is the creation of a large variety of bosons. The different types of bosons are created in the successive stages of the cascade. The mass of particles, therefore, is acquired in stages, or increments. This aspect of the proposed scenario is also supported by Lederman's reasoning, who stated that a particle acquires "Higgs-generated mass increments".(7)

The theory of the Higgs field, however, posed one problem that has not been solved.  When the mass-generating effects of a Higgs field were calculated and applied to the most common elementary particle, the electron, Lederman said, the results were "embarrassing infinities". The infinities, he explained, "came from this: simply described, when one calculated the value of certain properties of the electron, the answer (...) came out 'infinite'. Not just big, *infinite*. (...) Feynman and his colleagues proposed that whenever we see this dreaded infinity appearing, we in effect bypass it by inserting the known mass of the electron. In the real world, we would call this fudging. In the world of theory, the word is 'renormalization', a mathematically consistent method of circumventing the embarrassing infinities (...). Thus, the problem of mass was bypassed - but not solved - and remained behind as a quietly ticking time bomb (...).''**(8)**

The crux of the problem, as interpreted by the Physics of Encounter, is that the "final" encounter in the cascade of encounters is not really a final encounter. There is no such event in the quantum vacuum fluctuations of the universe. The expansion of VE-surfaces continues without interruption. Each time an encounter occurs, a new QV-bubble instantly expands from the point of encounter. One VE-sphere after another expands from the infinitely many points of encounter in the universe. Each point corresponds to what physicists call *zero-point energy* (ZPE), which is the vacuum energy that pervades the universe.

The theory of oscillating strings resolves the infinity problem. As explained by the physicist Gabriele Veneziano, "strings abhor infinities. They cannot collapse into an infinitesimal point, thereby preventing the paradoxes associated with that concept. They limit (...) the calculated values of physical effects, which either tend toward infinity or toward zero in the conventional theories.''**(9)**

Lederman wrote a book entitled "The God Particle" about the efforts of eminent physicists to understand the nature of the force encapsulated in the elusive boson that they see as the key to overcoming the "embarrassing infinities". Seven years later, in the year 2000, the *International Herald Tribune* reported that physicists had come "tantalizingly close" to identifying the particle that "may be the source of all mass in the universe and is viewed as the missing link in explaining the laws of particle physics and nature itself".**(10)** Lederman had good reasons for using an awe-inspiring name for a particle that may be "the missing link in explaining the laws of nature".

> *God never wrought miracle to convince atheism,*
> *because his ordinary works convince it.*
> Francis Bacon

"Having donated all of its energy to the creation of particles", Lederman explained in his book about the "God Particle", "the Higgs field retires temporarily, reappearing several times in various disguises in order to keep the mathematics consistent, suppress infinities, and supervise the increasing complexity as the forces and particles continue to differentiate. Here is the God Particle in all its splendor.''**(11)** As interpreted here, the Higgs field "retires temporarily" when the VE-surfaces that *converge* on point of encounter create a new VE-sphere. It consists of hemispheres that *diverge* from the point of encounter. The Higgs field "reappears" when either of the diverging hemispheres expands toward a point where it encounters, in morphic resonance, another VE-surface. The hemispheres, as I explained, expand into opposite directions of time. Both collapse when either one of them is encountered because a VE-sphere collapses in its entirety when it is encountered at any one point.

Why is Lederman's "God Particle", in contrast to the other bosons, so elusive? Why is it so difficult to pinpoint its location? The Physics of Encounter suggests an explanation. Since the Higgs boson is created in a cascade of encounters that also creates "messenger particles", the Higgs boson must exist at the origin of the "messages". That origin is the vacuum energy in a VE-surface. It includes the quantum vacuum in the QV-bubble that contains the qualia of the observer's consciousness.

As explained in the preceding chapter, the "inner light" of consciousness exists in the nuclei of atoms interacting through the exchange of photons. A Higgs boson is a source of energy in an atomic nucleus, comparable to the quarks and gluons in the nucleonic plasma. A Higgs boson exists inside a field of inconceivably high energies. The CERN experiment was designed to create such energies by smashing particles containing quarks and gluons into its energy constituents. The "God Particle" is the energy that exists at the origin of the inner light of consciousness. That is the origin of an act of observation.

When an act observation contributes to the creation of an elementary particle, that event corresponds to what physicists call the collapse of a quantum mechanical wave function. As interpreted here, the wave function is the expanding surface of a VE-sphere. When anomalous events occur, the amount of energy involved in an act of observation is amplified by the hidden energy of the collective unconscious. I described that process in chapter eight. The process creates abnormally large VE-spheres. It is energized by the quantum vacuum fluctuations of space on which, as Puthoff said (7/2), "coherent patterns can be written".

When the surface of such a large VE-sphere is encountered, this releases a large amount of hidden energy at the point of encounter. Since the VE-sphere is abnormally large, that point may be quite far away from the observer. I described that process in chapter nine. The point is the location where a radar system may pick up the mass created by the anomalous event. The mass is the anomalous plasma of a UFO. The abnormally large amount of mass is created by the cascade of encounters at that location. The cascade is a Higgs Field.

The mass of a UFO will have an effect on a radar screen only if the radar screen is located at the boundary, or surface, of the abnormally large VE-sphere. The gravity effect of this mass is created by the abnormally large amount of vacuum energy in the VE-surface. The gravity effect exists only in the wave patterns of the VE-surface, which is continuously re-created while the UFO is being observed. The process is geometrically the same as the one that continuously creates the so-called *standing wave* of electrons inside an atom. If the radar screen is located outside the relatively narrow band of energy oscillations in the VE-surface, it will not be affected by the anomalous gravity field of the UFO.

The mass of a UFO exists as long as the oscillations of vacuum energy in the surface of the abnormally large VE-sphere remain stable. *Two symmetrical events* assure the temporary stability of the anomalous mass. Both are encounters between the surfaces of VE-spheres. Since the encounters occur in morphic resonance, they involve identical wave patterns. The symmetry, or *structural identity* of the wave patterns, exists at two locations. One location is the point where the VE-sphere originates, the other location is the point where the VE-sphere is encountered. Since a VE-sphere expands instantly, the two encounters occur at the same time.

The symmetry that creates mass at points of encounter between the surfaces of VE-spheres exists because VE-spheres consist of hemispheres that expand into opposite directions of time. They cause not only future events, but also *past events*. The mind of the observer participates in this process. This corresponds to the creation of physical reality in the "participatory universe" described by Wheeler. (7/10) The equations of physics that describe quantum mechanical events contain the *symmetry of causation* with respect to both directions of time. That symmetry is broken, however, by the effect of consciousness. For us, as we observe the reality that exists in the universe, time only flows forward. An event that occurs now and influences events in the past (backward causation) seems illogical. But in the mathematical equations of quantum mechanics, forward causation and backward causation are equally probable. They are symmetrical possibilities. The physicist Harald Atmanspacher, in a paper on causation and mind/matter interaction, quoted the Nobel Prize laureate Frank Wilczek, who pointed out that the "fundamental equations have the symmetry, but the stable solutions to these equations do not".(12) For the conscious mind experiencing itself and at the same time experiencing physical reality, time only flows in one direction: forward.

The mass of a UFO is created by *broken symmetry*. The disappearance of a UFO is also caused by broken symmetry. When a UFO is observed, the VE-spheres involved in the act of observation suddenly become abnormally large. When a UFO disappears, sometimes "like a light bulb that is switched off", and sometimes gradually fading away, the VE-spheres are reduced to their normal size. The energy that breaks the symmetry, in both cases, is the impact of a mind event that interrupts an ongoing process. The collective unconscious that keeps a UFO in existence while it is being observed no longer contributes abnormally large amounts of energy if the observer "snaps out of" of the trance-like state of mind that was created by the observation. A sudden stimulus that puts the observer back in touch with normal physical reality can do the trick. It is like the snap of the fingers or the clap of the hands that therapists use at the beginning and at the end of the hypnotic trance they induce in their clients to retrieve buried memories.

The re-establishment of normal reality may occur, for example, when the mass of a UFO interacts with normal atmospheric events and this event dispels some of the initially compelling aspects of its mysterious appearance. The sudden awareness of some other event that has nothing to do with the observation of the UFO can re-open the mind to the all of the normal sensory input.

Mass provides "staying power", a resistance to a change in motion. The science journalist Overbye compared the Higgs field, which creates mass, to molasses. It slows the speed of particles that move through it. The molasses, in this metaphorical image, is the cascade of encounters in the space that separates the surfaces of VE-spheres expanding toward the point where they encounter each other. The "molasses" gets thicker as the distance between the two VE-surfaces decreases. The QV-bubbles involved in the cascade of encounters become smaller. As their size approaches zero, the density of the mass approaches infinity. The "embarrassing infinities" mentioned by Lederman (10/8) are avoided in the scenario described in this chapter.

As long as the surfaces of two VE-spheres that originate at separate locations and expand toward a common point of encounter are continuously re-created, the process remains stable and symmetrical. The symmetry is broken when there is a jolt, a "final" encounter, and a VE-sphere of equal size *does not* expand from the point of encounter. When a UFO ceases to exist, this happens because the mind "snaps out" of the trance described above and no longer generates a succession of abnormally large VE-spheres. These VE-spheres had expanded toward the point where the mass of the UFO was created. The mind events that created the UFO were amplified by the fluctuating vacuum energy of the collective unconscious.

The scenario of broken symmetry is applicable to all elementary physical events when a certain amount of mass suddenly disappears from one particle and reappears in a different particle, or when mass suddenly appears as the characteristic of a newly created particle. Particles are constantly *transformed* from one "thing" to something else. The speed of this transformation varies. Physicists recognized this as a break in symmetry but could not explain why it occurs. In 2008, two Japanese physicists received the Nobel Prize for a theory that provides an explanation.

They introduced the so-called Kobayashi-Maskawa matrix of particle transformation. As interpreted here, the matrix corresponds to the configuration of VE-spheres interacting at points of encounter where mass is created or reduced. It is a process in which the observer participates. The speed of particle transformation varies because time is an observer-related effect.

The third recipient of the 2008 Nobel Prize in Physics, by the way, was Yoichiro Nambu, a string theorist. He argued that the thrusts of oscillating strings connect the so-called quarks in heavy particles.(13) Physicists have concluded that quarks account for the gravity effects of the particles that contain them. To probe into the ultimate mysteries of the structure of matter, heavy particles containing quarks are smashed in particle accelerators. The biggest one ever built for that purpose is at the European nuclear research facilities (CERN) in Switzerland. The concept of quarks is one of the issues I will take up in the following chapter.

<u>NOTES</u>
(1) Pilkington (2006), p. 83
(2) Ibid, pp.114 and 117
(3) Ibid, p. 132
(4) Ibid, p. 138
(5) Goldberg (2005), p. 10
(6) Lederman (1993), pp 367-369
(7) Ibid, p. 369
(8) Ibid, pp. 276 and 282
(9) Veneziano (2004), p. 37
(10) *International Herald Tribune* (2000), p. 5
(11) Lederman (1993), p. 398
(12) Atmanspacher *et al* (2006), p. 1
(13) Kast and Nestler (2008), p. 9

# Chapter 11

**The structure of the atom. Quarks and gluonic light. Splintering photons. UFOs and "solid lights". Echo waves and offer waves. Atoms marching in step.**

The particles that physicists, on a whim, decided to call "quarks", are part of the puzzle that the particle accelerator of the CERN laboratory near Geneva was designed to solve. They are assumed to be particles inside particles. They are either "up quarks" or "down quarks". The upward "direction", metaphorically speaking, decreases the mass of the particle containing quarks. The downward direction increases the mass.

Before we look at quarks, as interpreted by the Physics of Encounter, we must be clear about the two directions of time in the proposed scenario of interacting VE-spheres. It is important to keep in mind that the expansion of VE-spheres from points of encounter occurs instantly, without the elapse of time. Time is created, relative to the observer, by the encounters between the QV-bubbles in the surfaces of VE-spheres. QV-bubbles, as I explained, contain the qualia of consciousness. The encounters create new QV-bubbles, and the encounters that create the interactions of atoms in the brain assure a continuously changing mix of qualia.

The events at points of encounter create not only the subjective reality of consciousness, but also the physical reality in the world of elementary particles. The creation and decay of particles provides an objective criterion for the measurement of time. Since the collapse of a QV-bubble, due to an encounter, is instantly followed by the expansion of a new QV-bubble from the point of encounter, the "flow" of time, consisting of successively experienced qualia, occurs without interruption, in *continuity*. The benchmarks of particles, on the other hand, are specific, measurable amounts of mass and energy. Particles are quantum mechanical events occurring at interrelated points of encounter. These events are a *discontinuous* physical reality.

The "gaps" that characterize the discontinuity of quantum events, to paraphrase Zizek's metaphor (3/4), are filled with the hidden energy of the mind. Quantum mechanical reality exists in what I have called *QM-space*. That space is not the only reality in our universe. The other reality is the timeless reality of vacuum energy. It exists in what I have called *VE-space*. The two kinds of reality are *entangled*. This entanglement creates not only anomalous events. It creates all physical events. The degree of entanglement determines whether or not the event is "anomalous", i.e. deviates from what is normally observed. As described by the Physics of Encounter, matter and mind are entangled with each other and become an *emergent reality* through the quantum vacuum fluctuations of space, i.e. through the interaction of QV-bubbles located in the expanding surfaces of VE-spheres that encounter each other.

For reasons explained at the end of chapter three, time only flows "forward" for a conscious mind when the mind events are created in a brain that interacts with the physical reality outside the brain. Time flows into both directions, however, in the world of elementary particles. This means that the effect of a particle event occurring now can cause an event in the future or in the past. The "flow" of time into both directions, relative to what we experience as the present moment, occurs as the expanding surfaces of VE-spheres interact. A VE-sphere, as I explained, consists of hemispheres that expand into opposite directions of time relative to the point of encounter that created it. When VE-spheres are created outside a particular atom, some of these hemispheres expand toward that atom, others expand away from it. Since quantum vacuum fluctuations occur throughout space, a VE-sphere can also be created *inside* an atom. In that case, one hemisphere that expands away from its point of origin will have an effect *outside* the atom unless it is encountered and collapses before it leaves the atom. The other hemisphere expands deeper into the atom. For an observer of the atom, it will not exit "on the other side" of the atom because hemispheres expand into opposite directions of space and of *time*. The hemisphere that expands deeper into the atom, away from the observer, expands toward a point in time that lies in the observer's past.

Hemispheres *converge* on a point of encounter and then *diverge* from that point when the encounter has occurred. Time is a continuous process of divergence and convergence relative to points of encounter. To visualize the reality of time as a "dimension" that is not identical with any of the dimensions of space, we must avoid the intuitive conclusion that the effects going into an atom are transformed inside the atom, as described above, and then simply "rebound" back into the three dimensions of space. Then we would still just be dealing with space, not with time. Going one way in space, let's say "forward", and then reversing directions, going "backward", does not correspond to a the effects of forward causation and backward causation in the dimension of time.

The Physics of Encounter suggests that time "flows" through an atom in both directions, which means that all events inside the atom are influenced by the effects of forward causation and backward causation. The "flow" of time is the expansion of VE-surfaces toward or away from the point of encounter where the observer is located at a specific instant. The location of an observer in this scenario, as I explained, is the location of the QV-bubble that contains the observer's qualia of consciousness.

This interpretation of the "flow of time" provides the basis for an explanation of the anomalous phenomenon that observers of UFOs have described as *solid lights*. The odd experience occurs when the quantum mechanical events in the nuclei of atoms that create the changing qualia of the observer's consciousness (and with it, the experience of time) are temporarily interrupted.

Let us first look at the atomic nucleus and then at the other part of an atom, the electron shell that surrounds the nucleus and contains the so-called orbital electrons. The flow of time, as defined above, enters the nucleus of an atom from one side and exits the nucleus "on the other side". The entry point at the boundary of the nucleus is located in *QM-space*. The exit point on the other side of the nucleus is located in *VE-space*. The other side of the atomic nucleus is like the other side of the moon. We cannot see it "from where we are". Physicists have no measuring devices that can travel around the nucleus to investigate the reality on the "other side". There is no subatomic space travel.

The observer of an atom is, of course, located *outside* that atom, in the QM-space of quantum mechanical interactions. That space includes the brain and the body of the observer as well as the measuring devices needed to probe into the structure of matter. From that vantage point, a physicist can measure the effects of the atom as it interacts with physical reality. The measurements allow deductions about the quantum mechanical aspects of the nucleus. They create effects in QM-space.

The vacuum energy in VE-space also exists inside an atomic nucleus. In fact, the nucleus consists almost entirely of highly compacted vacuum energy. The physical reality of the nucleus at the center of an atom is a "mixed bag", so to speak. The atomic nucleus consists not only of elementary particles but also contains the basic ingredients of mind events: the qualia of the observer's consciousness.

> *A scientist is the way an atom*
> *understands another atom.*
> Ernesto Cardenal

Ernesto Cardenal, a theologian and a poet, was Minister of Culture in Nicaragua from 1979 to 1986. In his collected works *Cantico Cosmico* (The Cosmic Song) he spoke of a pulsating universe, of expansion and never-ending "collisions" continuously creating new events. The collisions correspond to the encounters between the expanding surfaces of VE-spheres that contain the power of the human mind. They are quantum vacuum fluctuations.

Encounters occur throughout the universe, in the void of outer space and in the highly compacted space of an atomic nucleus. The particles in the nucleus contain quarks. The Physics of Encounter suggests that quarks are created by VE-surfaces encountering each other inside the particles that owe their mass to such encounters. The surfaces, as I explained, contain the vacuum energy of QV-bubbles. I will use the term "VE-surface" for the expanding surfaces of QV-bubbles in the nucleus of an atom because the principle is the same as with the interacting surfaces of VE-spheres. In both cases, the expansion involves hemispheres expanding into opposite directions, either toward a hypothetical point of encounter with an observer or away from it. In an atomic nucleus, the hemisphere that expands "upward", toward the hypothetical observer of the particle containing quarks, transmits an up quark effect inside the particle, toward a point where the hemisphere encounters a down quark effect. The hemisphere that transmits a down quark effect expands "downward", deeper into the particle. The point of encounter with this hemisphere is located *farther away* from the observer in the dimension of time.

This scenario poses a conceptual challenge. We should not visualize the particle in an atomic nucleus, or the nucleus itself, as a three-dimensional blob of something-or-other. The dimension of *time* is also involved. The hemisphere that transmits an up quark effect transmits the effect of forward causation. Relative to an observer located outside the atom in which quark effects occur, the hemisphere that transmits a down quark effect transmits the effect of backward causation. Quark effects exist within a "timeline". An up quark effect carries the *possibility* of an event toward a point in time that is closer to the observer's present moment. A down quark effect transmits the effect of an event that has *already occurred* (relative to the observer). When that effect is transmitted into a particle, it increases the mass of that particle.

Things are different if we imagine the location of the observer to be inside the nucleus of the atom in which quark effects occur. I will get to that in chapter thirteen. The quarks in heavy particles, such as those in the particles of atomic nuclei, do not exist at points in space. They are points within segments of time. This is in keeping with the physical theories that describe quarks as spatially "non-localized events" within the heavy particles that contain them.

The scenario that a quark is an event that occurs at a point in time and is also the effect of an expanding surface is supported by experiments that were done at the Fermi National Accelerator Laboratory in Illinois. As reported by the science journalist Thomas de Padova, one of the physicists on the research team, Steve Geer, pointed out that a quark might not just be a point-particle, as had been believed. The results of the experiments suggested, Geer said, that a quark "might have a hard core at its center" as well as an "outer shell".(1) The outer shell, as interpreted here, is the expanding surface of a QV-bubble that creates an effect inside a particle containing quarks when the surface is encountered.

Up quark effects, in this scenario, do not end at the boundary of an atomic nucleus. They accumulate at the boundary until the accumulated energy enables them to "jump across" the electron shell into the space that surrounds the atom. The accumulated up quark effects become a *photon* that is emitted from the atom. In this way, the effects of up quarks originating in the nucleus of an atom influence the present moment of an observer located outside that atom. The observer experiences the effect of the photon at the instant the photon is absorbed into the atom that contains, within its nucleus, the qualia of the observer's consciousness.

The photon that is created through this process becomes part of a light wave. There are two kinds of light waves. They are called *offer waves* and *echo waves*. These waves are created through the effects of forward causation and backward causation, respectively. When a photon is *absorbed* into an atom and creates a measurable effect, it is part of an offer wave. When a photon is *emitted* from an atom, that event is not observable. It occurs before the observer's present moment, which is created by the *absorption* of a photon into an atom. The unobservable photon event (the emission) is part of an echo wave.

Echo waves, in the scenario proposed here, consist of photons that are created by the accumulated gravity effects of down quarks. The accumulation occurs in the *depth* of an atomic nucleus, at the "other end" of the timeline that extends to the observer's present moment. The photons that are created there also jump across the electron shell of the atom, but into the other direction of time. The accumulation of quark effects at the two ends of a time segment establishes the size of the photon effects created inside the nucleus of an atom. It is a process that I will not describe in detail here. Readers who would like to grapple with the more complex concepts used by physicists in this context are invited to thumb through Part Three of this book.

The two endpoints of the timeline that runs through an atomic nucleus correspond to the image proposed by string theorists. They are the endpoints of an extended energy string that consists of countless segments of synchronized oscillations. The string of synchronized energy effects corresponds to the gravity effect of a black hole, which I described in the preceding chapter. The vacuum energy in an atomic nucleus exists in black holes. The oscillating string segments that extend into the nucleus and beyond (into both directions of time) interconnect the quantum mechanical events in QM-space with the vacuum energy in VE-space. The quantum world of particles is energized by oscillating strings.

The structure of atoms mirrors the entanglement of the two kinds of reality in our universe, QM-space and VE-space. The space within which an atom is observed is QM-space. The space inside the nucleus of an atom is, as I said, mostly VE-space. In the electron shell that surrounds the atomic nucleus it is the other way around. The electron shell is the space where quantum mechanical events are created by interactions between the vacuum energy inside the atomic nucleus and the vacuum energy outside the atom. The electron shell consists of the *standing waves* that are electron orbits. The quantum mechanical events in the electron shell are the *quantum jumps* of electrons between those orbits. Standing waves, as I explained in the preceding chapter, are temporarily stable energy configurations. Quantum jumps are instantaneous changes in those configurations.

Quantum jumps characterize the *discontinuity* of quantum mechanical events in QM-space. Stable configurations of oscillating vacuum energy characterize the *continuity* in VE-space. The vacuum energy in VE-space exists in what the physicist R. A. Aranov called "a space beyond the boundary of classical space and of quantum mechanics".(2) VE-space corresponds to what Puthoff called the "pre-manifest reality" of quantum vacuum fluctuations. The fluctuations (oscillations) create observable events. Aranov suggested that "unity between continuity and discontinuity exists at the boundary between qualitatively different spacetime regions."(3) The mathematician Michael Manthey coined the term "co-boundary operator" in his description of a "buffer-event mechanism" that links the two different types of space.(4)

As interpreted here, Aranov's "boundary" and Manthey's "co-boundary operator" can be equated with the electron shell of an atom. It provides the "buffer-event mechanism" that connects the events in the nucleus of an atom with the events outside the atom. The electron shell of an atom is impacted by events that originate outside the atom and inside its nucleus. Impacts from outside the atom occur when the atom absorbs a photon. Impacts from inside the nucleus occur when the effects of up quarks or down quarks accumulate at either end of the timeline through the nucleus.

The accumulating effects inside an atomic nucleus can be visualized as waves that gradually build up until the accumulated energy is released as a photon. Before that happens, some of the energy waves "splash" into the electron shell and influence the orbital electrons. The energy waves spread into the electron shell and cause the quantum jumps of electrons. Similar events happen when the electron shell is impacted from outside the atom. This occurs when the atom absorbs a photon.

The wave aspect of a photon, in the scenario proposed by the Physics of Encounter, is the expanding surface of a VE-sphere. When a photon impacts on the outer electron orbit of an atom, this is an encounter between two VE-surfaces. The standing waves that are electron orbits are continuously re-created surfaces of VE-spheres. Since all encounters between VE-surfaces involve a cascade of encounters between the QV-bubbles within those surfaces, the absorption of a photon into an atom also creates such a cascade.

Inside the electron shell, the cascade of encounters "shakes up" the energy patterns of electron orbits and causes the quantum jumps of electrons. The energy of the photon "splinters", so to speak, as it is absorbed into the atom. A portion of this energy enters the nucleus of the atom in the form of oscillating strings. It becomes the hidden energy of down quarks, which carry the effect of events that have already occurred deeper into the atomic nucleus. Down quarks increase the mass of particles that contain them. The increase in mass corresponds to the energy of a quantum jump, of the "jolt" that I described in the preceding chapter.

The physicist von Eickstedt argued that the nucleus of an atom is a "flow of energy" that is "not a reality in space", but "something other than an earthly, three-dimensional core".(5)   This corresponds to the image I described above and supports the scenario that the nucleus contains the "non-earthly" reality of consciousness.   That reality is the "inner light" created by the absorption of photons into interacting atoms.

The inner light of consciousness is created by the down quark effects in atomic nuclei that interact through the exchange of photons.   The accumulation of down quark effects, as I explained, culminates in the emission of photons that become part of *echo waves*. These light waves are like light reflected by a mirror.  They are the effects of events created in the nucleus of an atom.  That light is produced by gluons, which physicists have described as the inner-atomic equivalent of photons. The *offer waves* of light are the light of photons that are absorbed into an atom.  The inner light is what I will call *gluonic light*.  It is the light of subjective experience, of an image that exists in the mind. The light that I will call *photonic light* is the objective reality created by the photons that impact on atoms.  Photonic light exists as measurable energy in QM-space.   Gluonic light exists as a non-measurable reality in VE-space.  It is the oscillating vacuum energy in an atomic nucleus.

The process that occurs in a light wave assures the normal correspondence between objective and subjective reality.  The echo waves produced by the gluonic light in an atom are absorbed as photonic light into a different atom, and so on in a continuous, two-way process involving the flow and counterflow of time in offer waves and echo waves, respectively. In normal states of consciousness, there is no abrupt, radical change in what we experience as reality.   There is agreement between what we experience and what other people experience under the same circumstances.

The normal process that connects the inner world of gluonic light with the outer world of matter and photonic light is interrupted when an outside event is observed that the mind does not instantly recognize as a normal event.  The mind dwells on that event if it appears to be meaningful and important.  In such cases, the effect of the collective unconscious may influence the mind through morphic resonance, as described in chapter eight.  This happens, for example, when a UFO is observed and the anomalous focus of the mind on that event produces a trance, an altered state of mind that is like a self-induced hypnosis.

During a trance, the outer world is shut out for a certain period of time. The mind is focused on the images produced by gluonic light in the atoms of the brain. Normally, those images would "mirror" outside events. During anomalous experiences, however, the input from photonic light may stop abruptly. The result is that the image in the mind is suddenly no longer an event during which things happen. The image becomes a static snapshot of the last normally experienced moment. Under ordinary circumstances, the absorption of photonic light energizes the creation of "live" images in the atoms of the brain. When this energy input stops, the changeover to a purely gluonic production of images results in a brief interruption of the energy flow. It is as though an electric light briefly goes out and comes on again during a switch from one energy source to another.

. When such a switch occurs in the brain, the mind sees what is like a *freeze-frame image* of an outside event. The term goes back to the time when "moving pictures" were created by film projectors that illuminated a screen with a succession of images called frames. Freeze-frame images can be used to highlight a detail of an event, for example in a sports report on television. For a comment on the perfect body control of a diver jumping off a diving board, a sports reporter may refer to a videotape recording of the dive and use a freeze-frame image to stop the motion of the diver halfway between the diving board and the pool. The viewer sees the motion and then, suddenly, the diver is seen suspended in mid-air.

This kind of snapshot of an ongoing event is seen by the mind during an anomalous event that involves what UFO researchers have called the observation of a "*solid light*". A beam of light seems to come out of a UFO like a brightly illuminated, lengthening rod. The lengthening suddenly stops in mid-air. The beam of light does not "arrive".

An encounter with a "solid light" emitted by a UFO was reported by the Russian fighter pilot Lev Vyatkin. It occurred during one of his training flights in 1967. The incident was briefly mentioned in the press of what was then the communist-controlled Soviet Union. After the fall of communism, Vyatkin wrote a detailed article about his experience in a UFO magazine. The English version of his text quoted here is from the translation published in the magazine. Vyatkin described the UFO as very large, and oval-shaped. "I saw the object when I looked up from my instruments. (...) A strange object so close to my plane could not help but worry me so I requested the Flight Commander Major Musatov at once: 'Who is in the zone?' He (...) answered to my surprise that there was nobody in the zone. (...) I considered my next move and then decided to make the left turn I had planned, trying to be as careful as possible. Hardly had I banked the plane to the left and adjusted the speed and thrust when I saw a flash of bright light from above straight on the course of my plane. Then a slanting milky-white ray appeared in front of my plane. The ray was closing in on my plane. Had I not leveled out, the plane would have run into the ray with the fuselage or, to be more exact, with the cockpit."

"All the same, I hit the ray with the left wing. I was approaching the ray at very high speed, not taking my eyes off it, so I had time to notice and feel something very strange. No sooner had the wing touched the ray than the latter broke into a myriad of tiny sparkles like those you see in a spent firework. The plane shook violently and the instruments read off the scale. 'What's the matter? Is the ray solid?' I thought instinctively, with my eyes still on the strange sparkling pillar which stretched downwards. Soon the light above and the ray below disappeared. (...) My flight ended safely. For many days afterwards the surface of the wing which had come into contact with the strange ray shone at nights as if to remind me of the phenomenon."

"One knows from experience that there is no such thing as 'solid rays'. And yet it was me, a pilot, who happened to encounter the impossible phenomenon in real life. I felt some relief when the popular newspaper 'Komsomolskaya Prawda' published an article entitled 'Cosmic Ghosts' (October 17, 1989). The article stated that 'solid rays' really existed and I had not been the only person to come into contact with them. V. Selyavkin, the police chief of Voronezh, Russia, described a similar experience he had when he found himself at night on a road in the town suburbs. 'Suddenly a ray of light fell down on me from above. It was so bright and powerful that I felt it physically. You won't believe it, but it pinned me to the ground with its weight. Then the ray moved aside and disappeared. I will not forget it as long as I live (...).' Many other sources mention this phenomenon so characteristic of UFOs. Another feature of the ray is no less strange. It can project itself from a UFO like a telescope support or a probe. It terminates abruptly."(6)

The police chief mentioned in Vyatkin's article experienced the enormous power output of light emitted by UFOs that I described in chapter nine. The "weight" of anomalous light that is experienced as a "solid light" and pushed the police chief to the ground corresponds to the gravity effect that instantly flattens the stalks in the farm fields where *crop circles* appear overnight. These anomalous sources of light also create strong magnetic effects. They influenced the instruments of Vyatkin's plane which, as phrased in the translation of his article, "read off the scale". The left wing of his plane, which "shone at nights" for several days after the incident, and the malfunction of his instruments, were evidence that he flew through an abnormally strong magnetic field. These are real physical effects, not imagined events and will be described in the next chapter.

Seeing a beam of light that projects itself "like a probe" and then "terminates abruptly" in mid-air, as described by Vyatkin, is the freeze-frame phenomenon I mentioned above. A momentary trance shuts out the normal photonic light, so that the gluonic light in the mind of the observer dominates the experience. Vyatkin was briefly out of touch with reality. The effect that stops the normal flow of time can be preceded or followed by a *slow-motion effect*. For filmed or videotaped events, that effect is produced by filming or recording a larger than usual number of successive images per second (like rapid snapshots of an ongoing event), and then projecting or showing the normal number of images per second. When Vyatkin flew his plane through an anomalous beam of light and saw that the beam "broke into a myriad of tiny sparkles like those you see in a spent firework", he experienced a slow-motion effect.

The proposed explanation of that effect is based on the scenario that photons "splinter" when they are absorbed into atoms. The splinters are the oscillating strings that create the qualia of consciousness in the quantum vacuum of QV-bubbles. Each of these QV-bubbles is like a pixel in the gluonic image created in an atomic nucleus and experienced by the mind. A pixel is the basic unit or picture element of modern video technology. The gluonic image that exists during an anomalous mind event is created while the mind remains focused on an event that occurred at a specific instant. The mind focuses on what it perceived as a mysterious event. While this focus persists, physical events continue to occur outside the mind.

The anomalous source of light continues to affect the atoms of the brain but these effects are temporarily prevented from reaching the interrelated atomic nuclei that contain the gluonic image. The effects accumulate outside the boundary of atomic nuclei. When the floodgates of sensory input are open again, an abnormally large number of pixels is "projected" into the mind. This corresponds to the slow-motion projection of images from a reel of film or a video tape. The abnormally high number of pixels "stretches" the perceived unfolding of an event by showing more details per unit of experienced time. A mind that has lost contact with outside reality has no way of knowing that the experienced moments of "inner time" differ from the normal flow of time that is experienced by a mind when it is in touch with outside reality.

The Physics of Encounter describes how the elapse of time can be slowed or speeded up by processes that influence the exchange of energy between events outside an atom and events in the nucleus of an atom. The exchange takes place across the electron shell of an atom.

Experiments clarifying the structure of atoms were done by Wolfgang Ketterle. He cooled down atoms to temperatures near absolute zero. To understand the process examined by Ketterle, the distinction between *microcosmic* and *macrocosmic* realities is important. Elementary particles are a microcosmic reality. Atoms are a macrocosmic reality. Particles are packets of energy. Atoms are the substance of matter. Newspapers reported extensively on Ketterle's experiments with super-cooled atoms when he received the Nobel Prize in 2001. As described by Axel Görlitz and Tilman Pfau, his experiments show that at such low temperatures, atoms "lose their identity as separate macrocosmic entities".(7) They become what Ulrich Schnabel and Max Rauner called a "hybrid between microcosmic and macrocosmic reality".(8) Rainer Scharf explained that "separate atoms start behaving like quantum mechanical entities and 'condense' into a single wave".(9) The super-cooled atoms are "united in a wave", Ketterle said in an interview, "and begin to march in step".(10)

Atoms that "start behaving like quantum mechanical entities", as Scharf explained, behave like particles, like packets of energy. When super-cooled atoms condense into a single wave, substance is transformed into energy. Ketterle's experiments reduced atoms to what they basically are: energy waves. The different kinds of waves inside an atom are what Jahn and Dunne (3/8) called "complexly intertwined potentialities". Starkly reducing the temperature of atoms unravels the intertwined waves of energy. The wave patterns are united in a single process, so that the energy of this process is close to the *zero-point energy* (ZPE) that exists as hidden energy throughout space. Zero-point energy exists at a temperature of absolute zero. That temperature cannot be created in physical experiments, only approximated. The concept of "absolute zero" is a mathematical abstraction, like the concept of infinity.

Zero-point energy is the vacuum energy in VE-surfaces. It exists in the quantum vacuum of QV-bubbles. The expansion and collapse of

QV-bubbles in the surfaces of VE-spheres are the wave patterns of the quantum vacuum fluctuations that pervade all of space as hidden energy. The expansion of VE-spheres toward points of encounter with each other is a timeless process. It occurs instantly. The Physics of Encounter describes the dialectical process through which timeless, hidden events create time.

In Ketterle's experiments, the size of atoms was reduced to almost zero. Temperatures affect the size of atoms. In chapter nine, I explained that the abnormally large power output of the light emitted by UFOs occurs when the size of atoms fluctuates. The fluctuation affects the size of the electron shell that surrounds the nucleus of an atom. The distance between the boundary of the nucleus and the boundary of the atom is, in alternation, abnormally small and abnormally large. The Physics of Encounter suggests that the discrepancy between the inner experience of time and the outer reality of time occurs when the outer reality consists of such anomalous fluctuations. These fluctuations influence, in turn, the frequency of quantum jumps inside the electron shells of the atoms in the brain.

The elapse of time experienced by the human mind is influenced by the influx of energy from outside an atom (QM-space) into an atomic nucleus (VE-space). The influx is transmitted through the electron shells of atoms, as described above. An anomalous mind event that remains focused on a gluonic image created inside an atomic nucleus may temporarily block out the energy influx of the photons that impact on the electron shell. The normal elapse of time is halted. If the anomalous experience is only a partial trance, some of the outside energy passes through the filter of the mind. Experienced time may be slowed or speeded up, relative to the normal elapse of time. When we dream, for example, the adventures we experience may seem to continue for hours. But experiments have established that dreams are over in a matter of minutes, at most.

Additional information about the concept of time, with reference to the physical reality that creates normal as well as anomalous mind events, will be presented in the following chapters. They clarify the reality of mental images, their transformation and perceived motion through the landscape of our mind. They also deal with subject matter that I touched upon only briefly in the preceding chapters, such as the healing powers of the mind and the physical processes that explain how the mind leaves the body during Near-Death-Experiences.

NOTES

(1) de Padova (1996), p. 32
(2) Aranov (1971), p. 270
(3) Ibid, p. 271
(4) Manthey (2006), pp. 8-9 and 21-22
(5) von Eickstedt (1954), p. 8
(6) Vyatkin (1993), pp. 21-23
(7) Görlitz and Pfau (2001), p. 51
(8) Schnabel and Rauner (2001), p. 41
(9) Scharf (2001), p. 47
(10) Ketterle (2001), p. 288

## Chapter 12

**Magnetism, spin, and the "handshake" between waves. Cold fusion and hot plasma. Archetypes and elusive particles. The magnetic effects of UFOs and Poltergeist events.**

The mind influences matter through quantum vacuum fluctuations, i.e. through vacuum energy. That energy is also called zero-point energy because it corresponds to the energy that exists at a temperature of absolute zero. Nothing observable or measurable exists at that hypothetical temperature, which is actually the *absence* of temperature and of all "things" that might be hot or cold, or possess any observable characteristics.

Vacuum energy is, nonetheless, an immensely fruitful source of *possible events*. This energy is, physically, "not yet anything", but it is the source that creates not only a breathtaking array of different particles, but also the basic forces of nature that physicists can measure, like magnetism or electricity. It also creates the properties of particles that physicists can deduce from what they observe, like particle spin. Through processes described in this book, the human mind can "vibe" with this energy. We can accomplish amazing feats if we "put our mind to it".

The non-observable energy harnessed by the mind, for example, can change the amount of light that passes through transparent glass. This was demonstrated in an experiment by Garret Moddel and Kevin Walsh. As interpreted here, this influence falls into the category of psychokinesis, defined as the ability to influence objects or events by thought processes. The Physics of Encounter suggests that psychokinesis influences the motion of elementary particles, and that all psychokinetic effects on objects and events result from influences on the interactions between the particles inside atoms. This changes the inner structure of atoms.

Moddel and Walsh interpreted the results of their experiment as comparable to the influence of a specific intention on the output of random number generators. Of the light that impacts on transparent glass, they pointed out, about 92% is transmitted through the glass and about 8% is reflected. "Considering the light as photons to be detected, one cannot know in advance whether any given photon will be among the 92% transmitted or among the 8% that are reflected. The reflection process is therefore a fundamentally random binary process, akin to a random stream of bits (...)."**(1)**

A "binary process" is a yes-or-no process that answers a hypothetical question: will this event occur or will it not? The event, in this context, is the detection of a photon, either a photon that has passed through the glass or a reflected photon. Each event is a "bit", which is an elementary piece of information. The process is a *random* process, meaning: whether or not the specific event occurs (let's say the reflection of a photon) is a matter of pure chance. The experiment showed that the mind can influence, through psychokinesis, the average percentage of transmission and reflection.

A *binary* process involves *quantum mechanical events*. The "stream of bits" mentioned by Moddel and Walsh is not a continuous stream. Each "bit" is a quantum mechanical event that either occurs or does not occur. The yes-or-no process involves a *discontinuity* of separate events. A *stream of mind events*, on the other hand, occurs in *continuity*. There is no interruption between the moments experienced in a conscious mind that is aware of its own existence. The qualia of consciousness blend into one another and exist as a vibrant whole that consists of many different sensory experiences, as well as thoughts, feelings, and memories. This includes the *intentions* that can influence, through psychokinesis, the quantum mechanical structure of matter.

Quantum mechanical events occur in *QM-space*. Mind events occur in the vacuum energy of *VE-space*. Vacuum energy exists in an atomic nucleus. It creates what I have called the "*gluonic light*" experienced by a conscious mind. The "*photonic light*" outside the mind consists of quantum mechanical events. They occur when atoms absorb photons. The *emission* of photons from an atom is energized by the atomic nucleus. The *absorption* of photons into an atom replenishes the vacuum energy in the atomic nucleus. The electron shell is the intermediary.

In the scenario proposed here, the electron orbits that surround the atomic nucleus in an electron shell are standing waves that are created at points of encounter between the surfaces of VE-spheres expanding *away* from the nucleus and the surfaces of VE-spheres expanding *toward* the nucleus. The "orbits", or standing waves, are VE-surfaces. The expansion occurs in the dimension of time and transmits the effects of forward causation and backward causation. As I explained in the preceding chapter, VE-surfaces "splinter" into oscillating strings when they enter the electron shell. The interaction of strings creates QV-bubbles inside the electron shell. QV-bubbles expand into VE-spheres. I described the process in chapter three. The wave patterns inside an electron shell are complex because the electron orbits in an electron shell are created not only by encounters between the surfaces of VE-spheres that originate *outside* the electron shell. Many of the VE-surfaces that are electron orbits originate *inside* the electron shell. They are the surfaces of VE-spheres created at points of encounter between VE-surfaces that have already expanded into the electron shell from opposite directions of time.

The electrons in the orbits, or standing waves, that surround the atomic nucleus are created at points of encounter but immediately cease to exist as point-particles because the points expand into QV-bubbles, and these become VE-spheres. The stable location of an electron in a certain orbit is assured only if another electron with opposite spin is located in the same orbit. The reason for this is not important in this context. What is important here is the concept of spin.

When a particle "spins", does it actually rotate? In chapter two, I already said that the concept of spin is only a metaphor. And yet, there seems to be some force in the universe that causes actual rotation on a grand scale. The science author Claus Peter Simon provided details: "We are rotating -without becoming dizzy- at a speed of about 900 kilometers per hour around the axis of our earth, our planet rotates around the sun at more than 100,000 km/h, and our solar system rotates at 800,000 km/h around the center of the milky way. What a strange universe."(2)

At the elementary level of reality, particle spin is a key concept. In the entanglement of the two kinds of reality that exist in our universe, spin literally plays a pivotal role. As interpreted here, spin is an influence on the "turn of events" that is just as real as the quantum jump of electrons that disappear from one "orbit" and instantly reappear elsewhere in the electron shell, shedding or acquiring energy in the process. Spin, which is also called *angular momentum,* is created by the vacuum energy in the surfaces of VE-spheres. It literally "makes the world go round". That is not just a play on words, as this chapter will show.

Most physicists, however, insist that the concept of spin does not refer to "energy" as defined by the mathematics describing the transformation of mass into energy, and vice-versa. Critics have pointed out that this prevents an adequate explanation of the evidence that there are some nuclear processes during which "more power is coming out than going in". The astounding power output of the light emitted by UFOs, which I described in chapter nine, is a case in point. The energy puzzle has been tackled by

*Cold Fusion Research.* The fusion is assumed to be a process through which the components of the quark-gluon plasma in atomic nuclei interact and become a unified source of energy. Proponents of this theory argue that particle spin in the plasma accounts for the abnormally large power output that occurs without the release of a corresponding amount of high-energy particles or radiation.

The dispute between the proponents and the critics of the cold fusion theory revolves around the creation and the annihilation of particles. An electron and a positron annihilate (destroy) each other in an explosive process that creates a photon. The opposite electric charges of the particles and the particles themselves disappear. The photon appears in a flash of energy at the location where the two particles encountered and annihilated each other.

This process is reversible. An electron and a positron appear at the location where a photon ceases to exist. According to the physicist Donald Hotson, an adherent of the cold fusion theory, this violates the "conservation of mass-energy", which is stoutly defended by most physicists. Hotson sees this as an indication that the concept of spin needs to be adequately integrated into the prevalent theories of mass and energy. The conservation of mass-energy is the principle that mass can be transformed into an equivalent amount of energy, and vice-versa, but that the total amount of both does not change.

An introductory text to an article by Hotson in the journal *Infinite Energy* states that "his professors taught that conservation of mass-energy is the never-violated, rock-solid foundation of all physics", and also taught that "a photon (...) 'creates' an electron-positron pair. (...) But the 'created' electron and positron both have spin (angular momentum) energy (...). By any assumption as to the size of electron or positron, this is far more energy than that supplied by the photon at 'creation'. 'Isn't angular momentum energy?' he asked a professor. 'Of course it is (...).' 'Then where does all this energy come from?' (...) 'We regard spin angular momentum as an inherent property of electron and positron (...).' 'But if it's real energy, where does it come from? Does the Energy Fairy step in and proclaim a miracle every time 'creation' is invoked, billions of times a second? How does this fit your never-violated conservation?' 'Inherent property means we don't talk about it, and you won't either if you want to pass this course'."(3) This tongue-in-cheek introduction, obviously written by Hotson himself, highlights not only the issue but also the difficulty of making a dent in the wisdom of mainstream science.

In his article, Hotson addresses the controversy about the concept of *potential energy*. This concept is often interpreted in terms of so-called virtual events. They are possible events that may or may not occur. If and when they do, potential energy becomes actual energy. The relation between virtual and actual events, Hotson argued, should not be regarded as a "loan" of energy from one type of reality to the other. "In what form does a 'relation' loan out 'pure energy'? Cash, check, or money order?" Even though each event exists only very briefly, Hotson said, "this amounts to a *permanent* loan of *infinite* energy. 'Creation' is the proper term for it: only God could have that much energy to loan. (...) *Somehow the 'created' electron has something like sixteen times more energy than the photon that created it.* (...) Spin energy is *real* energy. It is the angular momentum needed by the electron to set up a stable standing wave around the proton. Thus, it alone is directly responsible for the extension and the stability of all matter."**(4)** (Italics by Hotson)

What Hotson called the "extension" of matter refers to the specific size of particles and atoms. They take up space. Energy, on the other hand, exists in a "field" without boundaries in the expanse of space, or at a point (where a photon is absorbed into an atom, for example). Hotson also referred to the standing wave around a proton. A proton is the positively charged particle in an atomic nucleus. In the interaction between the nucleus and the electron shell, a proton is the important particle because of the attractive force between the proton's positive charge and the negative charge of an electron. A standing wave in an atom, as I explained, is the "orbit" of an electron.

Hotson's arguments support the scenario proposed by the Physics of Encounter. The spin energy of particles is "real" energy. It is just as real as the particles. Particles are created by the interaction between VE-surfaces. The same goes for the creation of spin. I will describe that process below. Hotson sarcastically rejects the notion that spin is created by an "Energy Fairy" making transient "loans" of energy, "billions of times a second". This argument, however, misses the point. Interacting VE-surfaces contain oscillating vacuum energy. It is the hidden energy in VE-space. The measurable energy of quantum mechanical events exists in QM-space. VE-space "loans out" energy to create events in QM-space. The two kinds of reality in our universe are entangled. Oscillating strings connect the entangled realities. Energy is injected into QM-space at one endpoint of an oscillating string and returns into VE-space at the other endpoint.

The proposed scenario of the process that creates particle spin is based on the metaphorical description of a photon as an expanding VE-sphere. Photons are the energy of light waves, which consist of VE-spheres expanding and collapsing at points of encounter. The wave patterns in the surfaces of VE-spheres correspond to what physicists call wave functions. The term refers to the mathematics that allow physicists to calculate the probability of event. In this case, it is the probability that an event will be caused by a photon at a certain location. This scenario is in keeping with the way the physicist John Cramer (9/7) described a photon: "A photon emitted from a source has no directional preference. The emission creates a spherical wave function that expands like an inflating bubble."

The bubble, in this scenario, expands into a VE-sphere. When atoms exchange photons, the surfaces of these VE-spheres expand toward each other and meet halfway between the atoms that are interacting through the exchange of photons. The surfaces are part of the offer waves and echo waves described in the preceding chapter. They are hemispheres traveling into opposite directions of time. The two surfaces expanding toward each other contain wave patterns, or oscillating thrusts, created by the expansion and collapse of the QV-bubbles within these surfaces. The thrusts transmit effects into all three dimensions of space. They are the thrusts of oscillating strings.

The interaction between these hemispheres does not cause their collapse. This is only the case when the thrusts cancel one another in all three dimensions of space at a point of encounter, as explained in chapter three. When offer waves and echo waves interact, the encounter between the thrusts oscillating within one surface and the thrusts oscillating within the other surface is not a "head-on" encounter. There is a minimal time difference between the thrusts directed toward each other in each of the three dimensions of space. The thrusts are like the parallel motions of two smooth surfaces gliding past each other in *opposite directions* while the distance between them is minimal, just above zero. The surfaces rub against each other, so to speak, but move on as the hemispheres continue to expand and pass through each other.

This "rub" is the energy of spin. Imagine the smallest possible particle located between two such surfaces, with one surface on the left, moving upward, and the other surface on the right, moving downward. The surfaces would rub against the particle and cause it to spin clockwise. If the surface on the left moves downward and the other surface moves upward, the particle would spin counter-clockwise.

As I explained in chapter two about the use of metaphors in physics, spin is not the actual rotation of a particle. It is the "rub" that changes the observed characteristic of a surface. The "spin number" refers to number of observations that must occur before the same kind of surface is observed again.

When the oscillating thrusts in interacting surfaces rub against each other and the surfaces move on, the surfaces *permeate* each other. They pass through each other and continue to expand toward the respective atoms that will absorb their effect. The two surfaces are the spherical surfaces of photons. While they pass through each other, countless rubs re-create the spin effect. The surfaces are the wave fronts of light waves. The process corresponds to what John Cramer called a "handshake" between wave fronts "in which momentum, angular momentum and other conserved quantities are transferred".**(5)** The "conserved quantities" are the characteristics that particles acquire at locations where expanding VE-surfaces interact. One of them is angular momentum, or spin.

Momentum is like the thrust of a car speeding along a straight road. Angular momentum is the thrust of a car traveling around a curve. Spin is angular momentum because all VE-surfaces are curved surfaces. As described in chapter three, the oscillating string segments that create and connect the points in the VE-surface of a newly created QV-bubble carry a thrust "full circle". The same process occurs while QV-bubbles expand into VE-spheres. There is no limit to their size. VE-spheres expand until they are encountered. On a cosmic scale, oscillating strings are the force in the universe that causes the rotation of planets and solar systems, as described by Claus Peter Simon (12/2).

The oscillations in VE-surfaces permeating each other create not only the effect of particle spin. They also create the effect of magnetism. Spin is created when the thrusts are directed *toward* each other, with a minimal separation between them so that they "rub against each other" but do not encounter each other "head-on". A *magnetic line of force* is created when the oscillations are parallel thrusts into the *same direction*. By themselves, the oscillations are only *potential* energy in the expanding surfaces of

VE-spheres. They add up to the *real* energy of a magnetic line of force when they occur in parallel and with minimal separation. While the expanding surfaces of VE-spheres continue to permeate each other, the combined effects of the parallel thrusts establish the strength of the magnetic field. A magnetic field becomes permanent through the continued exchange of photons.

The surfaces in this scenario are the *hemispheres* of VE-spheres that have expanded toward each other. Their interaction begins when the respective forward points of the two hemispheres reach the location where they become one and the same point. This location is the same point in *space*, but not the same location in *time*. When the two hemispheres begin to interact, the two points are superposed in the dimension of time. The distance between them is minimal. It corresponds to the length of an oscillating string.

The two hemispheres continue to expand and to interact at minimally separated points of mutual permeation until both collapse when they impact on the atoms that absorb their effects. The hemispheres are the offer waves and echo waves of radiated light, which is electromagnetic radiation. The *magnetic effect* of light is created by the continuing "rub" of effects described above. The *electric charge* associated with light waves is created when VE-surfaces impact on each other in a specific configuration at a specific instant in time. (Part Three of this book explains the details of this process.)

Magnetism is what physicists call a *non-quantized* force. Its strength varies, depending on how many endpoints of oscillating strings are superposed in the dimension of time when the offer waves and echo waves of light interact. There is no encounter, only a *permeation* of wave patterns expanding toward each other. An electric charge, on the other hand, is a specific and invariable configuration of *encounters*. It is, therefore, a *quantized* force. The size of its effect remains constant.

The above scenario explains why strong magnetic effects are often associated with anomalous mind events. The effects occur when abnormally large VE-spheres interact. The spheres become abnormally large when the focus of the mind on a specific event temporarily shuts out the effects of other sensory input. Since all events create expanding VE-spheres, reducing the number of experienced events reduces the number of encounters that cause the collapse of VE-spheres. This allows the VE-spheres to expand into a larger size, picking up additional amounts of vacuum energy while doing so. This hidden energy becomes observable energy when the surfaces of VE-spheres interact in the configurations described above.

The effect that is created when the mind is emotionally focused on a specific event has been repeatedly recorded in Roger Nelson's Global Consciousness Project (8/2). The same type of effect occurs when an emotionally disturbed mind subconsciously focuses on "substitute objects" and causes Poltergeist effects, as described by William Roll (6/7). Another example of this effect is the magnetic force associated with UFOs. As I explained in chapter eight, the focus of the mind on a seemingly mysterious event in the atmosphere can trigger the release of hidden energy stored in the collective unconscious. Where UFOs are observed, strong magnetic fields influence electrical systems. Automobile engines stall. The compass needles in airplanes gyrate crazily. In the crop formations associated with the appearance of UFOs and patterned by the collective unconscious, the batteries of the equipment used by investigators fail. When the participants in séances conjure up phantom images, refrigerators stop working and lights flicker, to mention just a few of the magnetic effects that occur during anomalous events.

Magnetism can nullify, or at least reduce, the effect of gravity. Magnetic trains are an illustration of the principle involved. If two metallic objects are magnetized with the appropriate polarity, they repel each other. If the wheels of the train and the rail on which the train travels are magnetized in that way, this reduces the gravity effect of the train. There is less friction and therefore less energy is needed to propel the train forward.

In the proposed scenario of Poltergeist effects, or Recurrent Spontaneous Psychokinesis (RSPK), the object that is lifted and then propelled through the air is magnetized by the focus of the mind on that object. The object does not have to be metallic. Every object reflects light, and the oscillating energy of light waves has a magnetic component. The same is true for the shelf on which an object rested before it was lifted by the RSPK effect. Normally, these magnetic effects oscillate in unison and are not at odds with the gravity effect that keeps the object in place on the shelf. This is changed if the object is subconsciously "stared at".

Sheldrake's experiments showed (5/8) that the "extended mind" can influence the atoms of another person's brain. People can sense when they are being stared at. Schwartz and Russek showed (5/10) that merely *imagining* to be looking at someone can create that effect. During Poltergeist events, the subconscious focus of the mind on an object influences the atoms of that object. The magnetic component of the light reflected by the object is increased. The magnetic effect of the shelf on which the object rests, however, remains unchanged. The discrepancy between the two magnetic fields results in a force of repulsion like the one described above.

When magnetic polarities counteract the effect of gravity, even a heavy object can become an easy prey for the oscillating thrusts that a disturbed mind subconsciously focuses on the object. William Roll, a physicist who investigated and personally witnessed many poltergeist occurrences, described the "most likely scenario" for poltergeist phenomena as a "suspension of gravity at the site". He pointed out that a similar argument about the effect of an "RSPK agent" (the person who triggers Poltergeist events) was made by Puthoff, "who proposed that an object may be freed from gravity/inertia if the RSPK agent affects the zero-point energy (ZPE), a sea of random electromagnetic fluctuations that fills all of space. The agent would not generate the energy for object-movements, but would (...) loosen the hold of gravity/inertia that ordinarily keeps things in place."(6)

This supports the scenario proposed by the Physics of Encounter. The energy for the movement of objects is provided by the hidden energy of the collective unconscious. That is the oscillating vacuum energy in the interacting surfaces of VE-spheres. Vacuum energy, as I explained, corresponds to what physicists call zero-point energy.

A number of physicists have acknowledged the possibility that the quantum vacuum fluctuations of space, as a field of zero-point energy (ZPE), might correspond to the global consciousness field described by Nelson, or to the collective unconscious described by Jung. As Roll pointed out, Puthoff "suggests that if the ZPE is involved in RSPK, this shows that the ZPE has a consciousness component".(7)

In the scenario proposed by the Physics of Encounter, the mind of the person who triggers Poltergeist events does not have to provide a large amount of energy to cause the movement of objects. Once the gravity effect is eliminated, if only for a brief moment, very little energy is needed to "flick" the object off the shelf on which it was resting. In some instances, however, described by Roll, the Poltergeist effect thrusts heavy objects over distances of several feet. That's when the hidden energy in our universe really kicks in.

The concept of hidden energy makes many physicists uncomfortable. The same goes, to a lesser degree, for the concept of *potential energy*. What is this non-measurable "something" that is not-yet-anything? A heavy object teetering at the edge of a cliff, for example, "contains" potential energy. If you give it a gentle poke, it will fall and impact on the ground with plenty of real energy. That is *kinetic energy*, the energy of motion. The large *mass* of the object corresponds to its potential energy, which in turn depends on where the object is located.

The potential energy exists while the object is falling. It exists, so it would seem, *inside* the mass, at the location of the falling object. But is that really so? There would be no such energy if space did not contain the mysterious force that makes objects "fall", or move toward one another: the force of gravity. A similar force of attraction exists, as potential energy, at the location of an electric charge. If an electron and a positron come near each other, their opposite electric charges create a force of attraction. As described by Hotson (12/3), the two particles annihilate each other in an explosive event that creates a photon, the energy of light.

The physicist Michael Ibison pointed out that the word potential can be used "to describe a state of affairs in the continuum of space induced by an electric charge or a gravitating mass (...). This potential by itself is not an energy, has no mass, and cannot be weighed. The potential at some location in space refers instead to the energy that *would* result if a ('test') charge or mass were present at that location. One might say that it is a 'potential *for* energy'."**(8)**

The potential for energy exists because our universe contains the mass of objects and the space in which these objects are located. The "emptiness" of space contains something: the effect of gravity. To create *real* energy, however, gravity has to make things move. There has to be the opposite of emptiness: the "fullness" of mass inside observable objects. Unobservable by our senses, however, there is an even greater fullness in the emptiness of space. That is the incredibly large amount of fluctuating vacuum energy.

Ervin Laszlo (5/12) described this energy as "the originating source of matter itself." The quantum vacuum contains a staggering density of energy. According to John Wheeler, Laszlo pointed out, the matter equivalent of this energy is more than all the matter in the universe put together. Vacuum energy is not a "substance", but it creates the effects to which our senses react. This corresponds to the process by which the vacuum energy in interacting VE-surfaces creates elementary particles and the measurable energy of magnetism and electric charge.

> *That without substance can enter where there is no room.*
> *Great fullness seems empty,*
> *Yet it cannot be exhausted.*
> Lao-Tse, Chinese philosopher, 6th century, B.C.

The inexhaustible source of hidden energy in the universe creates what Einstein called the "geometry of space" that determines how objects move, relative to one another, in accordance with the influence of gravity. This influence, as Einstein said, is not a "force" like the measurable physical forces. Gravity pervades space as an invisible reality. As interpreted by the Physics of Encounter, Einstein's "point-particles" that exist at locations where the influence of gravity can be mathematically localized are points of encounter between the expanding surfaces of VE-spheres. They are the points where the invisible reality of gravity creates the visible reality of light. The points are the origins of photons. A photon, which Cramer (9/7) described as an inflating bubble, is a QV-bubble that expands into VE-sphere. Gravity is the potential energy that exists in an expanding VE-surface. Since photons transmit the gravity effects of light waves, Cramer called the inflating bubble of vacuum energy "Einstein's Bubble".(9)

When VE-spheres expand into an abnormally large size, for reasons described above, their interaction creates anomalous mind events and abnormally strong gravity effects. The gravity effect can be abnormally strong, like the light of the UFO described by Vyatkin (11/6), which pressed a Russian police chief to the ground. The gravity effect of the mind, amplified by the hidden energy the collective unconscious, can affect other minds. It can transmit images, as shown in so-called "remote viewing" experiments. In these experiments, one person sees a mental image of what another person is looking at. The participants who acted as "receivers" of this anomalous influence were able to draw surprisingly accurate representations of what the "senders" were seeing.

The observed shapes of UFOs are created by the same type of process. By influencing Einstein's "geometry of space" it can also produce the geometric configurations of the so-called *crop circles* in farm fields. I mentioned the phenomenon in chapter eight, where I explained how the collective unconscious creates structures. The patterned gravity effects press the stems of the plants in the crop formations to the ground, but the stems are not broken because the anomalous influence changes the structure of the atoms in these stems. The process changes the bonding of the atoms, briefly "softening" the stems so that they can be readily bent by the anomalous gravity effects localized in the space above them. The bonding between the atoms is loosened by changing the configurations of the electric charges in the electron shells that interconnect the atoms. The process is called "ionization".

The same process allows the mind, through psychokinesis, to bend forks or spoons, as described by Pilkington (6/15). In contrast to normal ionization, the internal "softening" of the substance by the anomalous influence of the mind occurs during a fleeting moment only. Its duration corresponds to the number of points, superposed in the dimension of time, where the oscillating effects in VE-surfaces interact, as described above. Whether this type of process includes the alleged ability of paranormally "gifted" persons to make seamless rings interlock is open to debate, but in the scenario proposed here it is a theoretical possibility. One substance can, in principle, "move through" another substance if the latter is briefly changed so that it becomes like transparent glass that lets the energy of light pass through it.

The balls of light commonly called *ball lightning* seem to be the kind of substance that can affect solid substances in that way. According to a German newspaper, the British electronics professor Roger Jennison reported that, while he was traveling in an Eastern Airlines plane that was flying over New York City, he saw a bluish-white ball of light, about four inches in diameter, emerge through the closed cockpit door of the plane. The luminous ball moved very slowly, about 30 inches above the floor, along the edge of the aisle, toward the rear of the plane, where it exploded without causing any damage.(10) The often reported ability of ball lightning to pass through a solid substance like the closed window of a house, or to enter flying airplanes, seems to indicate that ball lightning can briefly change the inner structure of solid matter. The luminous ball maintains its shape, like a solid object, while traveling through another solid substance.

The structure of a substance is largely determined by the process that interconnects the atoms of that substance. The atoms are interconnected because they share some of the electrons in their electron shells. This is where the anomalous plasma of ball lightning and of UFOs is created. The process creates the strong power output of these luminous phenomena. I described the process in chapter nine. The plasma is created in the space between the interconnected atoms because the size of these atoms oscillates. In one instant, the atoms are abnormally large, and in the next instant they are abnormally small, in continuous alternation. The time difference between these instants is minimal. It corresponds to the oscillation of the energy thrusts in the surfaces of VE-spheres. These oscillations create the effects of magnetism and spin, as described above.

In contrast to the plasma in the nucleus of an atom, the anomalous "plasma" that is the substance of UFOs and ball lightning is not hot. It exists outside the atomic nucleus. This is why, amazingly, the enormous power output of the light emitted by UFOs and ball lightning is "cold" (no hotter than normal light). The theorists of cold fusion are right that this anomalous process produces a large amount of energy without producing a correspondingly large amount of high-energy particles or radiation. They are right that the process during which "more power is coming out than going in" involves the effect of particle spin. As explained above, the magnetism associated with the electromagnetic radiation of light is an equally important factor.

Since UFOs have anomalous magnetic effects, the frequency, characteristics, and location of UFO observations are influenced by a variety of normal physical forces, such as the seismic forces related to earthquakes and shifts in the structures of the earth's crusts, and interplanetary magnetic fields. Mark Rodeghier, for example, reported a correlation between the galactic cosmic ray count and the frequency of UFO sightings.(11) I already mentioned the high number of UFOs sighted over power plants and military installations. These observations are particularly often confirmed on radar screens. As interpreted here, this is due to the interaction between the abnormally strong magnetic fields of UFOs and the electromagnetic fields at the above-mentioned sites. The electromagnetic fields at these sites are of above-average strength because of the generators at power plants and the electronic equipment used by the military.

The processes that result in the observation of UFOs are the same as those investigated in Nelson's Global Consciousness Project (8/2). Both involve magnetic effects. Nelson's project showed that events which evoke strong emotions worldwide influence the output of random number generators now operating at about 60 locations in more than 20 countries. Hans Wendt found that the polarity changes in the interplanetary magnetic field are correlated with the occurrence of the events investigated by Nelson and "with the corresponding emotional states".(12)

The Physics of Encounter interprets Nelson's concept of a global consciousness field and Jung's concept of the collective unconscious in terms of an entanglement between the two types of reality that exist in our universe. One is the hidden reality of oscillating vacuum energy that exists in what I have called VE-space. The other reality is the measurable energy of quantum mechanical events that exists in what I have called QM-space. The entanglement produces anomalous physical events and anomalous images in the mind. UFOs are images that impose themselves on the mind by the effect of the collective unconscious.

Even more startling are the images called *archetypes*. They impose themselves on the mind when it encounters the "alien beings" perceived as the occupants of a UFO. Ervin Laszlo called these images "irrepresentables" (not adequately describable) and pointed out that the psychiatrist C.G. Jung introduced the term archetype to explain how processes in the mind are influenced by "the myths, legends and folktales of a variety of cultures at various periods of history". Jung discovered, Laszlo said, "that the individual records and the collective material contain common themes. This prompted him to postulate the existence of (...) the 'collective unconscious'. The dynamic principles that organize this material are the 'archetypes'. (...) Jung formulated his concept of the archetype in collaboration with Wolfgang Pauli. He was struck by he fact that while his own research into the human psyche led to an encounter with such 'irrepresentables' as the archetypes, research in quantum physics had likewise led to 'irrepresentables': the micro-particles of the physical universe, entities for which no complete description appeared possible."(13)

Particles are the building blocks of atoms. The processes described in this chapter involve the indescribable aspects of reality that change the inner structure of atoms.

> *The problems of language here are really serious.*
> *We wish to speak in some way about the structure of the atoms.*
> *(...) But we cannot speak about atoms in ordinary language.*
> Werner Heisenberg

The problems of language are conceptual problems that arise from what Laszlo described as the baffling interaction between events in the psychological world (the "psyche") and the physical world ("physis"). He pointed out that the influence of one type of event on the other seems to be "acausal", meaning: without a recognizable cause. "The single factor that underlies the irrepresentables", Laszlo wrote, "may be the same as that which underlies the synchronicities Jung had investigated: meaningful coincidences that tie together in an acausal connectedness the physical and the psychological worlds. (...) As Charles Card summarized, (...) 'archetypes act as (...) fundamental dynamical patterns (...) In the realm of the *psyche*, archetypes organize images and ideas. In the realm of *physis*, they organize the structure and transformations of matter and energy (...)'."**(14)**

The important point in Laszlo's paper is that archetypes, or in more general terms, the hidden energy patterns that exist in the collective unconscious, organize the structure and transformations of matter and energy. As interpreted by the Physics of Encounter, the collective unconscious is a source of energy in the quantum vacuum fluctuations of space. That energy is contained in the timeless reality of expanding and collapsing VE-spheres.

*NOTES*
(1) Moddel and Walsh (2007), p. 20 (Abstracts)
(2) Simon (2004), p. 3
(3) Hotson (2002), p. 43
(4) Ibid, p. 49
(5) Cramer (2006), conference presentation
(6) Roll (2003), p. 79
(7) Ibid, p. 81
(8) Ibison (2007), p. 573
(9) Cramer (2006), conference presentation
(10) *Bild-Zeitung* (1990), p. 5
(11) Rodeghier (2007), conference presentation
(12) Wendt (2007), personal communication
(13) Laszlo (1996), p. 6
(14) Ibid, p. 6

# *Chapter 13*

The invisible energy field associated with life. Can plants learn?
Healing and the information in morphogenetic fields. Weight changes
at the moment of death.

> *With all your science can you tell me how it is,*
> *and whence it is, that light comes into the soul?*
> Henry David Thoreau

The light that comes into the soul - consciousness - originates in the
quantum vacuum fluctuations of space. Consciousness and the free will of
the human mind are a force that vibrates in the atoms of matter that is alive:
in the atoms of the brain. But the human soul is part of a larger, cosmic
whole. It is an integral part of the miracle of life. This poses a broader
question: how does living matter differ from non-living matter? What is this
"life force" that distinguishes humans, animals, trees and flowers from the
reality of barren rocks and other dead matter?

We need to recognize that the force that created our universe is not a
barren, unfertile reality. Some 15 billion years ago, it exploded into an
unorganized mass of energy droplets, then solidified into atoms and
molecules, into planets, suns, and galaxies. The right mix of energies that
evolved on our planet brought forth an abundance of life forms.

The previous chapters have provided arguments for the assumption that
our universe has no beginning and no end. It continuously renews itself.
Matter and energy are eternal realities. But a physical system imbued with
the force of life exists for only a limited period of time. Death is inevitable.
Experiments have shown that when life leaves a physical body, the weight of
that body increases briefly, then stabilizes again. We can assume that the
force of life has something to do with the volatile, ever-changing mix of
matter and energy. This chapter suggests a scenario for the process that
continuously renews itself within a physical system as long as that system is
alive.

The suggested scenario is based on the assumption that our universe is
eternal and that all observable reality arises from an unobservable,
non-measurable reality: the quantum vacuum fluctuations of space. The
fluctuating vacuum energy is the force that created our universe and
continuously creates the black holes in which other universes are born. It is
the force that creates life. It is the source of mankind's collective

unconscious and of our individualized consciousness. It provides "the light that comes into the soul".

Our consciousness is individualized because it is determined, to a large degree, by the activity of our brain. But the brain does not dictate everything we do. That's because there are two realities in the universe: observable reality and the unobservable force that brought forth our universe. There are measurable *brain events* and non-measurable *mind events*. I already mentioned that the Nobel Prize physiologist John Eccles (7/3) described the mind as an independent reality that acts upon the brain "in accord with its attention and its interests and integrates its selection to give the unity of consciousness from moment to moment". Eccles went on to argue that the human mind does not only "read out selectively" from the ongoing activities of the brain "but also modifies these activities."**(1)** As interpreted here, the human mind, as an independent reality, is empowered by an unobservable, eternal energy source that exists throughout the universe. It is the source that endows mankind with a *free will*. It is the light that "comes into the soul".

This light, this energy, is created at the instant a biological cell is fertilized and begins to grow into a human being. The Physics of Encounter suggests that this unobservable "inner light" exists in all physical systems that are alive, not only in the brains of humans. Experiments have shown that plants possess an "awareness" that allows them to react to specific states of the human mind. The process is, in principle, the same as the one that enables the human mind to sense the mind events of another person. As the anthropologist Jeremy Narby put it, "plants don't have brains so much as act like them".**(2)** To cite just one example: plants can "hear" music. Judging by their rate of growth measured in experiments, they seem to prefer the music by Bach.

The science journalist Kim Ridley, who described Narby's research, pointed out that plants process information about their surroundings with the same mechanism we use in our brain cells.**(3)** Narby lived among the Indians in the western Amazon of South America to study their customs. He saw that the medicine men (shamans) drink a brew of herbal extracts at ceremonies during which they experience hallucinations. They told Narby, according to Ridley, that the brew "enables them to communicate with plant spirits, which often appear as a pair of intertwined snakes. They view these spirits as 'animate essences' that exist in all life forms." Narby's hunch, Ridley reported, is "that Amazonians who drink (...the...) brew, to learn the healing properties of plants, take their consciousness down to the molecular level to gain access to information related to DNA, (...which is...) the molecule common to all life, from bacteria to bananas to human beings".(4) Interestingly, the "intertwined snakes" are shaped like the intertwined threads of DNA molecules.

Plants, it seems, can "learn". Scientists at the state university in the republic of Kasakhstan experimented with a philodendron, which is a tropical vine. Their experiments showed that the plant learned to distinguish a rock containing metal ore from other rocks. The experimenters gave the philodendron an electric shock every time they placed a rock containing metal ore near the plant. It soon began to react with measurable changes in the flow of electricity within its leaves whenever that kind of rock was brought near it. The plant did not react that way to other rocks.(5) It reacted like an animal in similar experiments. If an animal is given an electric shock every time it makes "the wrong move", it learns to avoid that move. Plants, of course, stay in one place. The science author Gerd Schuster quoted the biologist Jack Schultz who explained: "Plants are like slow animals. The only thing they cannot do is run away."(6)

A learning experiment was done in Japan with a single-cell organism called slime mold. The living speck of slime learned how to travel through a maze to find what Kim Ridley called "its favourite snack (oatmeal)". Ridley reported that the anthropologist Narby, mentioned above, "interviewed Toshiyuki Nakagaki, a Japanese scientist who reported in the journal *Nature* that slime mould could repeatedly find the most efficient route through a maze to find that delicious oatmeal on the other side. In explaining this phenomenon, Nakagaki introduced Narby to the Japanese concept of *chi-sei*, which means 'the capacity to know' inherent in all forms of life. *Chi-sei* was an idea Narby had been searching for – a way to bridge modern science and ancient wisdom."(7)

Many scientists investigating this "capacity to know" have concluded that it is associated with what are called *"morphogenetic fields"*. The term morphogenetic is a combination of the Greek words "morphe", which means shape or form, and "genesis", which means the way in which something becomes a reality. The invisible energy of these fields determines the shape and capabilities of an organism that grows from a single biological cell. Unexpected data obtained in learning experiments strongly suggest the existence of such fields. Experiments at Harvard University in the 1920s seemed to show that animals can pass on to their offspring knowledge they have acquired through learning. Since this was totally "out of line" with the conventional wisdom of biologists, other scientists, notably in Melbourne, Australia, also got into the act. They, too, used rats in their experiments. The Melbourne scientists discovered something even more surprising. The knowledge acquired by their rats through learning was passed on to other rats of the same species that were not offspring and had not been given any opportunity to learn!

As described in *Phänomene*, a reference work on anomalous phenomena, the rats were placed into a water basin from which they could only escape by swimming to either one of two planks that led to firm ground. One plank was brightly illuminated, the other was dark. Rats that left the basin by running over the illuminated plank received an electric shock. The first generation of rats learned to avoid the illuminated plank after each rat had received about 160 electric shocks. The following generations learned more quickly. After 30 generations, only about 20 shocks were required for the learning process.(8) When the Melbourne team, after 25 years of additional experiments, published its final report in 1954, that report pointed to a profound mystery. Where does the knowledge of an animal come from if it is not passed on from one generation to the next and not acquired through learning?

The answer suggested here: the morphic resonance that allows configurations of unobservable energy to interact also transmits information within the morphogenetic fields shared by animals that are of the same species. Morphogenetic fields are quantum vacuum fluctuations on which, to use Puthoff's words, "coherent patterns" have been written. These patterns are part of the collective unconscious.

There are similarities between an invisible, non-measurable morphogenetic field and the known fields of physics: the gravitational field and the magnetic field. The invisible "shape" of a magnetic field becomes visible if iron filings are strewn around the two poles of a magnet. The magnet pulls the filings into a circular shape surrounding the two poles. A gravitational field "bends lightwaves" and determines, as Einstein said, "the geometry of space". As I explained, anomalous mind events can create abnormally strong gravity effects and abnormally strong magnetic effects. Through psychokinesis (PK), the invisible energy of the mind can influence the substance of a physical object or make the object move as desired. The mind does this subconsciously when it creates poltergeist events through RSPK (recurrent spontaneous psychokinesis). During poltergeist events, the weight, or normal gravity effect, of an object can be briefly cancelled by an anomalous magnetic effect that "lifts" the object away from the surface on which it was resting. This magnetic effect is the force of repulsion between opposite magnetic polarities.

What needs to be defined, within the framework of the above scenario, is the process that distinguishes biological (*living*) matter from inanimate (*non-living*) matter. The question, in terms of mind/matter interaction, is this: What are the physical processes associated with the mind events in a brain that is *alive*, and how do these processes change when a person dies and the brain becomes an inanimate physical object? In more general terms, we need to identify the physical process that creates the *invisible energy associated with life*. The morphogenetic field enables lower forms of life to regenerate limbs that were destroyed. Within limits, all biological matter repairs damaged structures. When the bodies of animals and humans are injured, the wounds heal. Even more fundamentally, the morphogenetic field enables a single biological cell to grow into a complex organism containing countless interacting cells. Morphogenetic fields create "structures" where none existed before.

The energy in morphogenetic fields counteracts the influence of *entropy*, but only for a certain period of time. Entropy eventually "undoes" what the force of life creates. Humans, animals, and plants die. Within limits, the medical profession can prolong the life of humans and animals, and protect their biological structures against an untimely death. Entropy is described by the second law of thermodynamics. It states that the order within a physical system inevitably disintegrates into disorder, unless energy is introduced into the system from an outside source. An outside source of energy is necessary to create order, or structure, where none existed before. Human beings, pitting their will and their skill against the second law of thermodynamics, have harnessed the potential energy within their minds by learning the laws of nature. They have created structures where none existed before. They have built engines and produced the fuel for these engines. The fuel is energy from an outside source, like the food that humans and animals need, or the substances in soil and water that plants need for their growth.

Life on our planet evolved over millions of years. What is the outside source of energy that fueled the creation of life? The question touches upon the mystery of existence. Why is there "*something*" instead of "*nothing*"? As explained above, the Physics of Encounter suggests that the universe has always existed. When our universe exploded from "nothing" at the instant of the big bang, this "nothing" was the *zero-point energy* of quantum vacuum fluctuations. The fluctuations are the unobservable source of outside energy contained in the expanding surfaces of VE-spheres that have *not yet encountered one another* in morphic resonance. In the universe that exists today, oscillating strings straddle two minimally separated locations: the location of a reality that is not-yet-something and the location of an elementary event. They are the locations of unobservable vacuum energy and the locations of energy *created and observed* at points of encounter between the surfaces of VE-spheres. At the points of encounter, particles are created, together with the ingredients of the conscious human mind.

Since the interaction between VE-spheres that creates consciousness also creates morphogenetic fields, all human minds can, in principle, focus the "light that comes into the soul" on the biological processes that sustain growth and assure a healthy development. This ability can become a *healing force* that benefits other persons if it is enhanced through morphic resonance. It releases additional amounts of *outside energy* at points of encounter between the surfaces of VE-spheres. I have already described the process that focuses the vacuum energy contained in VE-surfaces and influences the structure of matter through psychokinesis.

Healers do not have to know the specifics of the mechanism involved. The information is contained in the subconscious depths of their mind. The focus of their attention produces a strong empathy with the afflicted person. Healers use their ability intuitively. They tap into the energy and boundless reservoir of the collective unconscious. The process corresponds to the one described by the science journalist Kim Ridley (13/7), who pointed out that the Japanese have a word for it: *chi-sei*, the "capacity to know" that is inherent in all forms of life, even in the biological structures of plants.

The human mind, energized by the vacuum energy in the surfaces of interacting VE-spheres, can influence biological matter in precise accordance with a pre-stated intention. A good example of this was provided by experiments with DNA molecules. Healers have demonstrated that they can change the structure of these molecules. DNA is the substance that determines the process through which a biological cell becomes what it is "destined" to become. The blueprint for a future reality is, so to speak, built into one tiny DNA molecule. Human DNA contains information about countless characteristics, physical and psychological, that distinguish one person from another.

When healers intuitively tap into the boundless reservoir of the collective unconscious, they access the reality that has been described by people who had a *Near-Death-Experience* (NDE). In an interview reported by Christian Stahlhut (8/14), an architect described how he experienced this stunning influx of knowledge and understanding. "Suddenly, I understood a lot of things (...). I understood the entire galaxy, the entire earth as a living being, the whole solar system as a living being. (...) I understood the DNA code, back then, which was 12 years before Crick and Watson were awarded the Nobel Prize for what they discovered. (...) At the time, I understood everything and now everything is gone." This brief influx of abundant knowledge corresponds to the experience of *chi-sei*, the capacity to know that is inherent in what the architect called "the entire earth as a living being".

The DNA code, or "blueprint" for future biological events, is contained in a *DNA helix*. It consists of two intertwined spiraling threads of which each is like the thread that winds around an imaginary screw. Experiments done by Glen Rein have shown that conscious intention can influence a DNA molecule in a test tube, "causing it to either wind or unwind", as desired. The winding and unwinding, Rein explained, can be measured by changes in the absorption of light. He pointed out that the influence of the mind on DNA is like the energy used by healers. In his experiments, he used people who meditated as well as some gifted healers. The experiments showed that the mind can influence DNA from distant locations. The mind, Rein said, has a "component which is able to carry specific information over long distances. Of course it is well known that healers can project their energy and heal someone thousands of miles away."(9) This amazing ability can be explained by the scenario proposed here. The healing effects as well as the qualia of consciousness are contained in the instantly expanding surfaces of VE-spheres.

The source of energy for anomalous healing is often called a *biofield*, or "biologic field" (that's the right spelling). Beverly Rubik, an eminent American expert on alternative medicine, mentioned that S. Savva "considered the biofield to go beyond electromagnetism, involving a non-physical mental component that carries the information of intention (...)". She also mentioned W. Tiller, who "proposed the existence of a new force to explain certain features of life, in addition to the (...) known forces of physics". Rubik pointed out that many researchers regard the biofield "as a holistic or global organizing field of the organism (...). Similar to the way a holographic plate distributes information throughout the hologram, the biologic field conveys information throughout the organism and is central to its holistic integration."(10)

The energy of "*qi*", a concept in Chinese philosophy and medicine, is similar to a biofield. The "*qi* field" is believed to supply the healing energy when doctors who have turned to alternative methods of medicine use *acupuncture*. It is the practice of piercing certain parts of the body with needles to treat a disease or to relieve pain. This energy, Rubik explained, "travels along the acupuncture meridians to all organs and tissues of the body. (...) Science has failed to substantiate the acupuncture meridians anatomically. Instead, these may be energetic manifestations of the biofield. Moreover, acupoints have been shown to be special regions of higher electrical conductivity than the surrounding tissues."(11)

Rubik's explanations support the scenario suggested by the Physics of Encounter. *Acupoints* may be points where encounters occur, in morphic resonance, between the expanding surfaces of VE-spheres. Encounters create elementary particles and the electric charges associated with them. The acupuncture *meridians* may be the interconnected points where VE-surfaces pass through each other and create the effect of the left-handed particle spin associated biological molecules. I explained both processes in the preceding chapter. Acupuncture meridians cannot be substantiated anatomically because the surfaces of VE-spheres are an unobservable reality. They contain the potential energy that creates the measurable reality of elementary particles.

A hologram, mentioned by Rubik in her description of a biofield, is a three-dimensional holographic image produced by laser beams. It is like a ghostly image suspended in space. An observer attempting to touch a holographic image will grasp into nothing. The image seems to be a graspable, three-dimensional reality in front of the observer, but its origin is elsewhere. The origin is the holographic plate from which a laser beam "distributes information throughout the hologram", as Rubik put it. Each point in the holographic plate contains information relevant to the entire image. A (theoretical) point, by itself, is a zero-dimensional reality and therefore "nothing". Together, the interrelated points are "something": in image, or information.

The role of the holographic plate in producing a non-graspable image is similar to the role of the brain in producing qualia, the elementary ingredients of consciousness. If you close your eyes and imagine to be looking at something, what you see in your mind is a non-graspable image. During a Near-Death-Experience (NDE), the brain processes the equivalent of holographic images that contain a wealth of normally inaccessible information. The brain is temporarily in touch with a larger reality, in space and time, that exists outside the human body.

During an NDE, the brain is "dead" in the sense that all electrical impulses have ceased. But when the clinically dead person, whose heart has stopped beating, is finally revived, the brain once again begins to function normally. Because the brain had not succumbed to final death, and remained capable of being reactivated, it was able interact with the potential energy of the biofield, the force of life and consciousness that pervades all of space.

The next chapter will describe the process by which the brain, during an NDE, accesses abnormally large amounts of information contained in mankind's collective unconscious and experiences events that transcend the limitations of space and time. In the remaining part of this chapter, I will explain how the force of life that exists in biological matter is created by the vacuum energy in the interacting surfaces of VE-spheres.

Let me start by repeating two points. Matter and mind share a common source. Matter and energy are equivalent realities. But what is energy? As interpreted by the Physics of Encounter, the influence of the mind on the physical reality of matter is produced by what Ibison (12/9) called a "potential *for* energy". That is not the same as energy, and more precise than the term "potential energy". For simplicity, I will continue to use the latter term. The potential energy of the human mind is not an observable physical reality. The elementary particles of matter, on the other hand, are observable. They represent a measurable amount of energy. The Physics of Encounter describes how the potential energy in the surfaces of VE-spheres creates the observable energy of particles. The unobservable reservoir of energy exists in the quantum vacuum fluctuations of space, on which, as Puthoff said (7/2), "coherent patterns can be written".

What is the "coherent pattern" of matter that is *alive* (biological matter)? What distinguishes living matter from non-living (inanimate) matter? One of the distinguishing characteristics is the spin effect. In the molecules that are the main building blocks of biological matter (proteins), left-handed particle spin predominates. Physicists aren't sure why this is so. As suggested by Hotson (12/3) and other theorists of "cold fusion", particle spin provides the energy that is not accounted for by the equations of electromagnetic radiation, of interacting particles and physical forces. These equations provide adequate explanations for all events involving inanimate matter, but cannot account for the process that creates consciousness and matter that is alive.

The Physics of Encounter suggests that *chi sei*, the "ability to know" mentioned by Ridley (13/7), exists as potential energy not only in biological matter. It exists in the atoms and particles of all physical reality, in the rocks tumbling down a mountain cliff and in the ocean waves splashing against the hulls of ships. Like individual mind events, all matter events (including those that nobody sees) are part of the collective unconscious, of an all-encompassing mind that causes all events in our universe. Our own experiences, and their all-encompassing cause, are rooted in the laws of nature. Our experiences, our acts of observation, contribute to the creation of what we observe. Atoms and particles exist because we exist.

To understand the difference between living and non-living matter, we need to distinguish between the complex biological processes occurring today, after many years of evolution, and the hotly debated "spark" that triggered the beginning of life. It was, on a very much smaller scale, like the spark that triggered the big bang. The school of thought that supports the scenario proposed here has presented evidence that the onset of life occurred four billion years ago at the bottom of the ocean in towering accumulations of calcium carbonate. The geologist Mike Russell and the biophysicist Harold Morowitz have argued that the laws of nature made the evolution of life inevitable.

As reported by Manfred Dworschak in the newsmagazine *Der Spiegel*, hydrogen was continuously produced in the towering rock formations underneath the sea and was shielded, inside the rocks, from the carbon dioxide in the water. Alluding to the powerful chemical reactions that change the characteristics of interacting substances, Dworschak explained that the structures of the two different substances at the bottom of the ocean, shielded from each other, "harbored an urge to interact", and "searched for a way to release their pent-up energy". Dworschak compared the event that finally occurred to the flash of lightning in a thunderstorm.(12) The enduring flashes of newly created energies were the first steps toward carbon-based life-forms.

The continuous creation of pent-up energy at the subatomic level of reality, as interpreted here, is like an effect that favors the quantum jumps of electrons to a higher energy state inside atoms. This changes two interrelated factors: the *spin* of particles and the creation of *up quarks* inside the nuclei of the affected atoms. (Keep in mind what I said about quarks in chapter eleven.) An example of the appearance of something radically new from a continuous change in energy is what happens when the temperature of water is increased. At some specific point on the temperature scale, the liquid begins to turn into steam. In the philosophy of dialectical materialism, this is called the jump from *quantitative* changes (degrees of temperature) to a *qualitative* change: a new kind of reality.

In the creation of life from non-living matter, this jump is caused by the quantum vacuum fluctuations of space. The non-observable vacuum energy of these fluctuations exists in what I have called VE-space. The observable substance of matter particles exists in what I have called QM-space. The creation of the "ability to know", of consciousness, from the interaction of the particles in atoms is a quantitative change that changes the simplest types of matter to matter endowed with increasingly individualized structures of consciousness. As these quantitative changes influence one another, the process snowballs into what physicists call "self-generated complexity". The process that triggered the beginning of life is, in essence, the guiding principle of the complex biological processes that energize today's many life-forms-

In the following text, the term "observer" should be understood as the equivalent of all substances that possess the "ability to know", including the lowest living organisms. To simplify the language, I will present the evidence regarding the effects of particle spin and up quarks in terms of human consciousness. As explained in the preceding chapters, the energy of particles is determined by the configuration of points of encounter between the surfaces of VE-spheres. A particle is "observed" when the expanding surface of one of the interacting VE-spheres contains the elementary ingredients of the observer's mind event. The ingredients are the *qualia* of consciousness. They are created by the effects of oscillating strings in the quantum vacuum of the QV-bubbles that energize the expansion of VE-spheres.

As described in chapter twelve, the *spin effect is created in VE-space*, at locations where the expanding surfaces of VE-spheres, instead of encountering each other, pass through each other. The event corresponds to what Cramer (12/5) called a "handshake" between wave fronts. They are the *offer waves* and *echo waves* expanding toward each other in the space between atoms that are exchanging photons. Together, the atoms and the photons are an electromagnetic wave. In the scenario proposed by the Physics of Encounter, the *observer is part of the wave.*

The participation of an observer in an electromagnetic wave is twofold. The qualia of a particular observer's consciousness are contained in one of the two interacting surfaces that produce particle spin. When a photon is absorbed into an atom, the potential energy is transformed into the energy of what I have called the *gluonic light* inside the nucleus of the atom. The gluonic light becomes the "inner light" of human consciousness through the interaction of atoms inside the brain. As von Wijk explained (4/2), the atoms of the brain exchange *bio-photons.*

In chapter eleven, I already described how atoms emit and absorb photons. Let me nail down what is relevant here. The events that are caused inside an atomic nucleus by the absorption of a photon into the atom, and the subsequent transformation of the photon's energy into the gluonic light of human consciousness, must not be visualized as the motion of a physical object into any of the three dimensions of space. They are processes that create the experienced flow of time by continuously changing the qualia of consciousness.

This must be kept in mind for the following description of the particles called quarks. In the scenario proposed here, the effects of quarks and the quantum jumps or electrons are interrelated. Quarks exist inside other particles, including those in an atomic nucleus. There are up quarks and down quarks, but these terms have nothing to do with the dimensions of space. Quark effects, as I explained, must be understood as effects within the dimension of time. Relative to the present moment experienced by an observer, a quark may be the cause of an event that has already occurred, or of an event that will occur in the future. Down quark effects carry gravity effects (mass) deeper into the atomic nucleus, away from an observer outside the atom. Down quarks *increase* the mass of the nucleus. Up quark effects carry gravity effects toward the observer of the atom. Up quarks *decrease* the mass of the nucleus in the observed atom,

Before I come to the details of the processes involving quarks, let me mention the symbolic aspects of the "upward" force. It creates a higher level of reality. Like the corresponding quantum jumps of electrons, the upward force creates consciousness and life. In a metaphorical or spiritual sense, it counteracts the downward force that makes us the captives of gravity. It can carry our soul upward into the realm of weightlessness, where the soul, when it has left the body that has died, is freed from the burdensome weight of mass and the afflictions of a physical body. In a metaphor closer to physical reality, we might say that the upward force puts us in touch with the reality of a holographic image. The image exists in our mind when we close our eyes and imagine what we wish to see. The free will of the human mind symbolizes the freedom that exists where there is no mass, only energy.

The energy that is metaphorically called the "spin" of a particle fits into this scenario. The relationship between particle spin and the effect of quarks becomes apparent if we look at the spin and the electric charge of electrons and positrons. These particles are associated with the waves of light created by atoms exchanging photons. Electrons and positrons correspond to the effects of down quarks and up quarks, respectively. Electrons have a negative charge. Positrons have a positive charge. Positrons are the anti-particles of electrons. A positive charge carries a gravity effect away from the point of encounter that is the present moment experienced by an observer.

Positrons, as the Nobel Prize physicist Richard Feynman said, "travel backward in time". They travel away from the observer. Keeping in mind that there is no actual motion through space, only the instantly expanding surface of a VE-sphere, we can say that mass is created and observed when a VE-surface impacts on the observer, and that the observed mass is diminished when a VE-surface expands away from the observer in the dimension of time. That is why protons, which contain quarks and have a positive charge, weigh less than neutrons. Neutrons are exactly like protons except that the positive and negative electric charges of the quarks in neutrons balance out, so that neutrons are electrically "neutral".

In the scenario proposed by the Physics of Encounter, *a quark expands*. A quark is an effect inside a heavy particle. The effect is created at a point of encounter. The point expands into a QV-bubble which may, in turn, expand into a new VE-sphere. Research reported by de Padova (11/1) supports this scenario. The experiments suggested that a quark has an "outer shell". As interpreted here, that is the expanding surface of a QV-bubble. Under certain conditions, that surface can become one of the surfaces of the particle that contains the quark, or it can expand beyond the boundary of the particle. The details are described in Part Three of this book. Also important in this context is the fact that the spherical surfaces of quarks and of photons consist of hemispheres that expand into opposite directions of time. One hemisphere can create events in the observer's past, through backward causation, the other hemisphere can create events in the observer's future, through forward causation.

Photons are the accumulated gravity effects of quarks. As explained in chapter eleven, the effects of quarks accumulate in the nucleus of an atom and are sporadically ejected from the atom as photons. The explosive release of quark effects is, in principle, like the "pent-up" energy that triggered the creation of life. This scenario must take into account that an atomic nucleus, as described by von Eickstedt (11/5), is a "flow of energy" and "not a reality in space". It is, in his words, "something other than an earthly, three-dimensional core". As interpreted here, the boundary of a nucleus is not a spherical reality in space, but an oscillating, two-directional flow of energy in the dimension of time.

In the scenario proposed by the Physics of Encounter, time is created by the interaction of VE-spheres. Their spherical surfaces expand toward a point of encounter, where they create an effect, and then expand away from the point of encounter, in continuous alternation. This is the "flow" of time. The nucleus of an atom is an oscillating *segment of time*. The segment is like a *compass needle* within the nucleus, for which north and south correspond to the future and the past. When the event within the nucleus causes a future event, that is causation forward in time. When the effect of a future event impacts on the nucleus, that is causation backward in time. Events impact on an atomic nucleus from all directions of space. The "substance" of the nucleus consists of countless time segments, or "compass needles", pointing into all directions of space. The calculated diameter of an atomic nucleus is the spatial equivalent of a two-point segment of time. The boundary of the nucleus is blurred, or "fuzzy", because the time segments oscillate.

When the effects of down quarks accumulate as gravity effects in the nucleus of an atom, the accumulation corresponds to a segment of time consisting of superposed points of encounter. Relative to an observer of that atom, the final effect in the accumulation of down quark effects is the "deepest point" inside an atomic nucleus. That point is, paradoxically, not at the center of the nucleus, but at the boundary that faces away from the observer, like the unobservable backside of the moon.

The "deepest point" inside the nucleus is the point where the events in the nucleus create the gluonic light of consciousness. For one instant, that is the location of an observer. The observer is located where accumulated down quark effects are emitted as a photon. *The observer creates the photon.* When the photon created by accumulating down quark effects is emitted from an atom, the observer in that atom ceases to exist and is now located in the atom that absorbs the photon. The observer straddles interacting atoms because identical photons are instantly emitted and absorbed when atoms exchange photons. Photons are the surfaces of instantly expanding VE-spheres.

During normal mind events, the interacting atoms are located in a brain. The atoms interact by exchanging *biophotons.* During the anomalous mind events described in this book, the mind affects atoms outside the brain.

Now let's look at the effects of up quarks. They are effects that expand away from the observer's present moment (a gravity effect) that is created by the accumulated down quark effects in the nucleus of an atom. Up quark effects expand into the opposite direction of down quark effects. Up quarks are anti-gravity effects.

Photons are vacuum energy that accumulates in atomic nuclei. They are created by the down quark effects and the up quark effects that accumulate at opposite ends of the one-dimensional nucleonic time segment. In a quantum jump, the expanding VE-surface of a photon "jumps" across the electron shell of an atom and across space, impacting instantly on the atom that aborbs it. Recent research supports the assumption that no time elapses during this jump from one atom to the other. The photon emitted from an atom reduces the mass of that atom. The mass is re-created when the atom absorbs a photon. The exchange of photons between atoms causes minimal fluctuations of the gravity effects produced by the interaction of quarks in an atomic nucleus.

Normally, in the exchange of photons between atoms that are not alive, the up quark effects and the down quark effects created by photons (offer waves and echo waves) are in balance. This is not the case when the atoms of biological systems interact. The defining characteristic of such systems is the dominance of up quark effects and of particles with left-handed spin.

The interactions that are acts of observation are encounters between the QV-bubbles in the expanding surfaces of VE-spheres, of which one contains the qualia of the observer's consciousness. At the level of interacting atoms, acts of observation occur in all biological systems, not just in the human mind. When an atom is observed, the effect of a mind event influences the observed atom, and vice versa. The hidden energy of consciousness creates down quark effects in the nucleus of observed atoms. It creates up quark effects in the atomic nucleus that contains the qualia of the observer's consciousness.

Acts of observation are outgoing effects as well as incoming effects. Up quark effects are thrusts originating at the location of the observer within the nucleus of an atom and directed away from that location. They decrease the mass of the nucleus. Down quark effects are thrusts impacting on the nucleus of an atom. They increase the mass of the nucleus.

The photons that create the anti-gravity effects of up quarks in the nuclei of atoms are the echo waves of light. The photons that create the gravity effects of down quarks are the offer waves of light. Offer waves create mass. Echo waves reduce mass. Offer waves expand forward in time. Echo waves expand backward in time. The two types of waves pass through each other, continuously creating particle spin and magnetic lines of force.

The upshot of this scenario is that observed matter *weighs less* as long as it is alive, and that its observed mass briefly increases when death occurs. The occurrence of death ends the dominance of up quark effects. The weight soon returns to what it was before the moment of death because up quark effects are created through the backward causation of echo waves. The balance of up quark effects and down quark effects is restored when the down quark effects of the observed matter that has died are no longer counteracted by the dominance of up quark effects. Within a brief period of weight increase, the gravity effects of down quarks, created by the forward causation of offer waves, play a larger role, so that the opposite quark effects are brought back into balance.

Lewis E. Hollander has presented experimental evidence that supports this scenario. For his experiments, he used 12 farm animals destined to be destroyed. The animals were placed on a scale while they were still alive. A temporary weight gain between 18 and 780 grams, lasting up to six seconds, was registered on the scale at the moment of death of seven adult sheep. With the other animals, the results were not conclusive.**(13)**

Hollander mentioned experiments by other investigators allowing no clear-cut conclusions, but pointed out that spontaneous weight changes have been observed during dreams and meditation. He also mentioned experiments by Duncan MacDougall in 1906 who weighed humans at the moment of death, using a beam balance scale. His results seem to run counter to those reported by Holland, but the weight change observed by MacDougall, according to Hollander, were so "violent and abrupt" that MacDougall had to readjust his beam balance to compensate for the change.(14)

Reports of weight gains during *séances* were described by Rosemarie Pilkington. She mentioned the experiments by William Crawford, an engineer, who spontaneously decided to investigate the séances of a young girl named Kathleen. He had heard about the séances while lecturing on an engineering topic at a university in Belfast at the beginning of the 20th century. Crawford borrowed a large commercial scale for the occasion and found that, when Kathleen made a table rise from the floor, her weight increased.(15) Crawford found that the table offered an "elastic resistance" when he pushed it downward, but seemed to resist being pushed toward Kathleen. More about that in chapter fifteen.

Kathleen was also able to create an invisible influence that produced rapping sounds on a table. This type of anomalous event is often reported by participants in seánces. According to Pilkington, Crawford found that Kathleen's efforts resulted in a *weight loss*. "By having her sit on the weighing machine while the raps were produced, he found that Kathleen progressively got lighter. Her weight loss continued until the loudest, strongest blows were heard. At that point, her weight had dropped eight pounds but would then return to normal when the raps stopped." While the rapping sounds were occurring, Crawford reported, other participants in the séance also lost weight, but not as much.(16) I will describe how the mind can produce rapping sounds in chapter sixteen.

Weight changes also occur during Poltergeist events. In an article about such phenomena, William Roll pointed out that the weight of a person who was causing the events was measured during two episodes. According to Roll, the investigators, Hasted, Robertson, and Spinelli, reported in 1983 that the weight gain during each episode was about one kilo and lasted about five seconds.(17) The events that occurred when MacDougall measured weight changes at the moment of death were like Poltergeist events. They involved what Hollander (13/13) called "violent and abrupt" changes in weight. The beam balance scale used by MacDougall was thrown out of whack and had to be readjusted. The hidden influence on the scale showed, as Hollander put it, that "something dramatic occurred during the moment of death".

The processes that create both types of phenomena are similar. A temporary weight gain occurs when the hidden energy associated with living matter leaves the body at the moment of death. When people unknowingly cause poltergeist events, they use some of the hidden energy associated with the inner light of consciousness, amplified through morphic resonance with the collective unconscious. Investigators of poltergeist events call them RSPK (recurrent spontaneous psychokinesis) and refer to the people who cause these phenomena as RSPK agents.

As interpreted here, the energy used by an RSPK agent is an amplified anti-gravity effect of the positive charge transmitted by the unobservable echo waves of light. A positive charge, as I explained, is associated with the anti-matter particles called positrons. While the anti-gravity effects are being used up to energize poltergeist events, the gravity effects of matter play a correspondingly greater role, which increases the weight of the RSPK agent.

Death-related Poltergeist events were described by the pastor of the Memorial Church in the heart of Berlin, Wolfgang Kupsch, in a public lecture I attended in 1988. The experiences confided to him by members of his congregation who had lost loved ones, he said, were similar to what he himself had experienced at the age of seven. One night he and his mother were unable to sleep. He kept thinking of his grandmother. Suddenly, at ten minutes to ten in the evening, a huge flowerpot on a window sill tipped over and fell to the floor. The next day, they received a telegram informing them that the woman who was Wolfgang's grandmother had died. They later learned from relatives that the time of death was ten minutes to ten.

The cause of such events is a process that involves morphic resonance of the kind that triggered Shirley MacLaine's experience (5/11) when her close friend Peter Sellers died. Here we see the most striking evidence that physics can do more than help us understand the interplay between mass and energy. It can contribute to our understanding of life and death, and of the forces that create empathy, emotion, and love - the bonding forces of humanity.

*The day will come when, after harnessing the winds, the tides and gravitation, we shall harness for God the energies of love. And on that day, for the second time in the history of the world, man will have discovered fire.*
Teilhard de Chardin

*NOTES*

(1) Popper and Eccles (1977), pp. 361 and 514
(2) Ridley (2007), p. 28
(3) Ibid, pp. 30 and 32
(4) Ibid, pp. 30 and 32
(5) *Phänomene* (1994), p. 155
(6) Schuster (1989), p. 262
(7) Ridley (2007), pp. 28-29
(8) *Phänomene* (1994), pp. 343-344
(9) Rein (1995), pp. 173-180
(10) Rubik (2002), pp. 708-709
(11) Ibid, pp. 708 and 714
(12) Dworschak, p. 124
(13) Hollander (2001), pp. 495-500
(14) Ibid, p. 498
(15) Pilkington (2006), pp. 130-131
(16) Ibid, pp. 131-132
(17) Roll (2003), p. 79

## Chapter 14

**Near-Death-Experiences and holographic images. Stored information in a non-functioning brain. The out-of-body mind and the role of positronium.**

The remarkable aspect of a Near-Death-Experience is that just about everyone who has had such an experience no longer fears death. An NDE has profound spiritual consequences. It usually removes any nagging doubts about the possibility that our consciousness will survive the death of our physical body. The fervent hope of many people that there will be "a life after death" is a powerful emotional force that energizes religious beliefs. It is reconciled with the power of reason by the simplest of all "proofs" that have nothing to do with scientific reasoning: an NDE shows that self-awareness is very much alive, *outside the body*, while the body is clinically dead.

An NDE provides deep but transient insights into the nature of an intensely experienced reality that is not the graspable reality of material objects and human bodies. The insight is like the information contained in a *holographic image*, which appears to be a reality at a certain location even though nothing is actually there. The insight is lost when life is restored to the body. The information is too complex to be processed by the earthly computer that is our human brain. But the certainty remains: the experience does not contradict what our rational mind tells us about the reality that sustains our existence.

Our understanding of physical reality has been radically changed by what scientists call the quantum mechanical revolution. The seemingly hard and graspable reality of the physical objects that present themselves to our senses and interacts with our body are the product of interacting packets of energy, or elementary particles. Each packet, or quantum, of energy is, by itself, nothing graspable. It represents a measurable impact. As the mathematician Michael Manthey pointed out, our intuitive view of how material objects interact corresponds to "the universal image of billiard balls colliding and rebounding. (...) The quantum mechanical revolution put an end to this, evaporating 'materia' into a cloud of probability amplitudes (...).(1)

The scenario proposed by the Physics of Encounter is based on this new understanding of physical reality. The "probability amplitudes" mentioned by Manthey are the potential energy contained in the expanding surfaces of

VE-spheres. The surfaces, interacting at points of encounter, are the probability waves described by the theory of quantum mechanics. "Particles" are created at interrelated points of encounter between instantly expanding, invisible waves .

People brought back to life after an NDE said that the light they saw was like an image that evoked countless experiences at the same time. Christian Stahlhut is one of the many authors who have described "the other reality" of a Near-Death-Experience in terms of a holographic image. "It seems as if the difference between earthly reality and the other reality is only a difference of degree, not a fundamental difference. Both are constructs like a hologram (...), arising solely from the interaction between our consciousness and the world outside. In other words, our reality is something like a hardened version of the afterlife dimensions."(2)

Stahlhut pointed out that one of the central figures in NDE research, Kenneth Ring, "believes the holographic model to be the key to an understanding of Near-Death-Experiences". Ring, according to Stahlhut, "had noticed that most of the 'experiencers' he had interviewed told him that they were immersed in vibrating spheres, enveloped by harmonious music and brilliant colors, and attracted to a God-like light. Ring's explanation: Near-Death-Experiences are excursions into a realm where there is neither space nor time, filled with high-frequency vibrations that the mind translates into graphic images."(3)

> *Time and space are but physiological colors*
> *which the eye makes, but the Soul is light.*
> Ralph Waldo Emerson

The "high-frequency vibrations" in Ring's description of an NDE, as interpreted here, are the quantum vacuum fluctuations of space. The vacuum energy in these fluctuations is contained in the surfaces of VE-spheres interacting at points of encounter. The encounters correspond to the scenario described by the physicist Wolfgang Pauli (3/6). One surface contains the energy of the "self", expanding outward into the physical world, as Pauli said, the other surface contains the "onrush of non-self energy toward the self".

The points of encounter are like the points where a holographic image originates. The expanding surfaces of VE-spheres contain the qualia of consciousness. During an NDE, the mind sees the entire holographic image at a glance and experiences itself as a part of all of them. This includes not just the images of loved ones, but also the thoughts and feelings of people who are perceived to be present during this encounter in "the other world". During an NDE, the thoughts and feelings of others are experienced like one's own. That's because, as I explained, encounters occur only if there is a *morphic resonance* between the wave patterns of the interacting surfaces, which are probability waves.

Based on the scenario of interacting VE-spheres that create elementary particles at points of encounter, the Physics of Encounter proposes a description of the process that allows the human mind to experience itself hovering above its clinically dead body, from where it observes the successful efforts to restore life. The process that creates such an *out-of-body mind* involves what physicists call a *standing wave*. In an atom, such a wave corresponds to the "orbit" of an electron that hovers, so to speak, above the nucleus of the atom. As explained in chapter two, the word orbit is a metaphor. The electron does not actually circle around the nucleus. It is, rather, continuously re-created at various locations around the atomic nucleus. As interpreted here, those locations are points of encounter between the surfaces of VE-spheres that expand into the atom from opposite directions in the dimension of time. Those surfaces contain QV-bubbles, and the quantum vacuum in QV-bubbles provides the energy that causes the "quantum jumps" of electrons from one orbit to another.

Physicists have established that if two electrons have opposite spin, this allows both of them to stay within the same electron orbit. The opposite spin stabilizes the electron orbit. The Physics of Encounter suggests that an out-of-body mind event is created and stabilized in a standing wave that exists between atoms exchanging photons. During an NDE, a human mind is energized at an out-of-body location by the *backward causation* effects of echo waves containing the positive electric charge of positrons. Even after a brain has stopped functioning as living matter, it is, like all matter, still a source of both forward and backward causation effects. This follows from Wheeler's theory of a "participatory universe". As explained by Folger (7/9), Wheeler argued that ordinary, non-living matter can "transform what might happen into what does happen". In Wheeler's universe, Folger said, "clouds of uncertainty" that have not yet interacted with "some lump of inanimate matter" are part of the "vast arena containing realms where the past is not yet fixed". This applies mainly to the distant and largely unknown regions of the universe. But in the invisible arena containing the energy of life and of consciousness, the participatory universe plays an important role.

The "clouds of uncertainty" are VE-spheres interacting in the quantum vacuum fluctuations of space. Within the time frame established by the successful efforts to restore life, the brain, precariously close to being just a "lump of inanimate matter", remains capable of receiving and storing information. During an NDE, an out-of-body location of the conscious mind is established through the process described below. The unobservable reality at that location continues to interact with the clinically dead brain because, as the successful efforts to restore life continue, that information is transmitted, through backward causation, to the mind experiencing itself at the out-of-body location.

In the scenario proposed here, the out-of-body mind events occur within the VE-surface that is like a large, hemispherical dome above the location of the temporarily lifeless body. The hemispherical surface acts like a mirror that reflects the probability waves (i.e. the expanding surfaces of VE-spheres) originating at the location of the clinically dead body. When the out-of-body mind observes what is happening during an NDE, the process is like the one I described in chapter three. The qualia of consciousness are experienced *inside* the spherical surface that encloses the origin of the conscious experience.

The interaction of VE-spheres creates what I have called the *gluonic light* of consciousness in the nuclei of atoms. The gluonic light is experienced like a holographic image. The quantum mechanical processes that correspond to the interaction of VE-spheres create mass, electric charge, magnetism and spin at the location of the out-of-body mind. In contrast to the mind events rooted in the brain, the physical reality of an out-of-body mind contains very little of what Stahlhut (14/2) called the "hardened version of the afterlife dimensions". Having no quantum mechanical storage capacity, the physical reality at the out-of-body location mirrors what is happening and transmits the information back to the brain cells that will soon be brought back to life.

The clinically dead brain, imbedded in Wheeler's "participatory universe", dutifully stores the flow of information from the out-of-body mind, to which it remains connected through backward causation. When life is restored, the brain cells snap back into action and allow the revived mind to experience the stored information in its entirety. The experience is actually a "memory", but it is experienced like the observation of an ongoing event. This scenario suggests that an NDE is not what it appears to be. When the NDE occurs, the mind that experiences itself as an out-of-body reality is already energized again by the quantum events in the revived brain.

What the mind observes when it experiences itself outside the body is not the "live" action of observed medical procedures and other events, but more like a *recording* of events that have already occurred. The information in the "recording" is played back at fantastic speed, relative to the events that occur outside the mind during the brief interval between biological revival and the new start-up of normal mind events. During that interval, the self is not yet in touch with outside reality. It experiences, for the first time, the unfolding of the events that have already occurred. They were preserved "for later use", so to speak, like the images and sounds stored on a DVD. Since the revived brain that plays the "recording" is not yet interacting with outside reality, it does not realize that it is experiencing what is, in essence, a *memory*. It experiences the flow of events like "normal time". Time is "relative", as Einstein said, and the compression or stretching of time during anomalous experiences provides striking examples of the physical processes that create the experience of time. More about *time warps* in chapter seventeen.

The out-of-body mind events, involving a minimum of "hardened" quantum mechanical matter, occur at a distance from the brain that corresponds to the distance between a standing wave in a morphogenetic field and the quantum events in the brain that generate the invisible field. Through the effects of backward causation, the two locations stay connected when the brain stops to function, because the brain still has the potential energy that is needed for the restoration of life. At the location of the out-of-body mind, the echo waves from the brain that has temporarily stopped functioning encounter the offer waves from the atoms and molecules of the surrounding space. *Offer waves* transmit the potential energy of forward causation, *echo waves* transmit the effects of backward causation.

Echo waves, as interpreted by the Physics of Encounter, consist of photons that are created by the accumulated up quark effects in the nuclei of interacting atoms. I described the process in the preceding chapter, in which I explained that up quark effects predominate in the atomic nuclei of matter that is alive. During an NDE, these up quark effects occur in a clinically dead brain that still contains the potential energy for a restoration of life.

In *Near-Death-Experiences*, out-of-body mind events are created by offer waves and echo waves interacting within a morphogenetic field. The process is similar to the exchange of photons between atoms, which absorb and emit photons in alternation. But since the interactions at the location of an out-of-body mind involve only a minimum of "hardened" quantum mechanical reality, they have to be energized by an outside source. That source is the potential energy in the brain that is only temporarily dead. The energy it regains when it is revived is provided "in advance", so to speak, through backward causation.

The offer waves and echo waves interacting at the location of the out-of-body mind transmit the potential energy of electrons and positrons, respectively. Their opposite spin and opposite electric charge create what is called orthopositronium. In that energy configuration, the electron and the positron do not immediately annihilate (destroy) each other in a flash of energy. Instead, they first create a magnetic line of force. Orthopositronium then becomes parapositronium, followed by the mutual annihilation of the electron and the positron. The particles and their opposite charges disappear and their energy explodes into two or more photons. This scenario from the literature of theoretical physics supports the scenario proposed here. As described in chapter twelve, opposite spin effects can merge as parallel lines of thrust, which together form a magnetic line of force. Because the expanding surfaces of VE-spheres contain layers of expanding and collapsing QV-bubbles, the VE-surfaces passing through each other create spin effects and magnetic lines of force in continuous alternation.

This is how Donald L. Hotson described the beginning of the electron/positron interaction, the creation of positronium: "(...) when an electron approaches a positron, they don't just rush together and disappear. Instead, they approach until they are a distance apart that is (...) some 56,000 times the diameter of a proton (...and then...) they start to orbit around each other in the configuration called 'positronium'. (...) After orbiting each other in this pseudo-atom for a time that depends on whether their spins are parallel or opposed, they emit two or more photons that total all of their positive energy. After that they are no longer detectable and conventional wisdom says that their charges and spins have 'cancelled' and that they have 'annihilated' and are no more."(4)

Hotson's rejection of this conventional wisdom supports the explanation of an NDE presented here. The "pseudo-atom" he mentioned exists at the location of the out-of-body mind and does not simply disappear. During an NDE, continuously re-created orthopositronium energizes the mind at an out-of-body location. Once established at the boundary of the brain's morphogenetic field, the out-of-body mind is kept in existence by the backward causation effect of the clinically dead brain, which is temporarily like the "lump of matter" imbedded in Wheeler's "participatory universe".

The Physics of Encounter suggests that Wheeler's scenario of backward causation keeps the clinically dead brain *in touch with the future*, when it will be brought back to life. As long as that event exists as a possibility, probability waves expanding backward in time carry potential energy into the brain. Because of that energy, the clinically dead brain is more than just an ordinary "lump of matter". Through backward causation, it keeps the out-of-body mind in existence during an NDE.

The location of the out-of-body mind is directly *above* the clinically dead body because the echo waves of backward causation that expand away from the clinically dead brain contain the potential energy of *positrons*, which are anti-matter particles. Normally, as Cramer pointed out (9/7), the echo waves and offer waves of photons (Cramer's "inflating bubbles") have "no preferred direction". But that applies only to the exchange of photons between *inanimate* matter, when up quark effects and down quark effects are in balance. In morphogenetic fields, unobservable energy effects are directed toward very specific locations to maintain the structures of the living organism.

When the effects of positrons create positronium at the location of an OBE, these particles of *anti-matter* counteract the gravity effects of *matter*. Matter that is not alive has *inertia*, which means that if it is at rest, it will remain motionless, and if it is moving (like a falling object), it will continue to move in the same direction unless it is affected by some outside force. Living matter, on the other hand, is able to move on its own. The direction of the effects that are necessary to keep the out-of-body mind in existence is *opposite* to the effects of gravity, which keep a lifeless body pinned to the surface of the earth. The location of an out-of-body mind, therefore, is directly above the body.

The out-of-body mind, as I said, is like a mirror that instantly reflects the information it receives. Like a satellite above the earth, this mirror beams information to the atoms of the brain. Since the brain still contains the *potential for life*, it is still "tuned", in morphic resonance, to the information it receives. It is able to store the information and to experience this information as soon as it is brought back to life. People who had an NDE reported that they saw and heard everything while doctors were laboring to restore their life, and that their mind then experienced, among other events, encounters with the images of other humans as well as a mix of colors and musical sounds. They experienced no sensations of taste or smell. This consistently reported fact about NDEs supports the scenario proposed here.

It provides an explanation why taste and smell are not included in the spectrum of Near-Death-Experiences.

Here's the explanation: when we *see* or *hear* something, we receive information contained in *vibrational energy*. Light waves are electromagnetic vibrations involving atoms exchanging photons. Sound waves are vibrations involving the molecules of the air. In the vacuum of empty space, there are no sounds. But when information is transmitted in the normal human environment containing the air we breathe, vibrations occur. According to the mainstream theories of sensory perception, however, vibrational energy is *not* involved when we taste or smell something. The preferred theory is the so-called "lock-and-key model". It states that the sensations of taste and smell involve the "protein receptors" of a living organism. The molecules containing the substances that create the sensations of taste and smell, according to this theory, fit like a key into a protein receptor.

If this is so, an out-of-body mind cannot process and reflect the information of taste or smell because it is not a molecular reality that possesses protein receptors. It is not a living organism interacting with protein molecules. The positronium energizing an out-of-body mind can, however, react to vibrational energy. It can reflect that energy and pass it on "for storage and later use" by the brain that is only temporarily dead. The NDE scenario proposed here supports the lock-and-key model in a controversy that has not yet been resolved.

Patients who were revived after an NDE have often reported that their out-of-body mind, floating near the ceiling of the operating room, passed through the walls into another room and saw what was happening there. Can the out-of-body mind do this *willfully*? The above scenario suggests that this is not the case. The observations made during an OBE are reflected back to the substance of the brain that has not yet begun to function again. What an out-of-body mind sees, therefore, does not involve the willful acts of a functioning brain. When the out-of-body mind floats through walls, these changes in location presumably correspond to the random fluctuations of the probability waves that are the interacting surfaces of VE-spheres.

According to Christian Stahlhut, the probability of having an NDE when the heart stops after a heart attack is between 3o and 60 percent.**(5)** Among the factors that determine whether or not a Near-Death-Experience occurs is presumably the random mental focus immediately preceding the event that triggers an anomalous experience in some people but not in others. In chapter eight, I proposed an explanation why dozens and sometimes hundreds of people may see a UFO, or even several of them flying in formation, while many other people at the same location do not. The same type of unpredictable brain events may be a factor here. NDEs, like UFO experiences, are physical events structured by the potential energy of the collective unconscious.

In a scientific paper on quantum physics and brain processes, Max Tegmark presented evidence that our perceptual process produces the outside reality we see. That can be a UFO or the holographic image we see in a Near-Death-Experience. Our brain processes information, but Tegmark stressed that *the brain does not function like a computer*, and that a computer cannot "have", or produce, consciousness, not even the most advanced quantum computer.(6) Tegmarks arguments support the scenario proposed by the Physics of Encounter, which describes the reality we see in terms of the expanding surfaces of VE-spheres that contain the qualia of consciousess and create structures where none existed before. It is not possible to build a "conscious computer" that can do this.

The creation of actual structures from an unobservable "outside source of energy" (the quantum vacuum fluctuations of space) corresponds to what is called a *decrease in entropy*. The term "entropy" describes the principle that it takes energy to create structures and to keep them in existence. Without that energy, all structures will eventually decay. The structures can be buildings or bridges, or life-forms, or the structured experiences of consciousness.

Tegmark described the process that produces consciousness in terms of three factors: subject, object, and environment. This supports the scenario presented here. The subject is one of the two VE-spheres that interact at a point of encounter. The object is the other VE-sphere. The environment consists of any other VE-sphere that may or may not come into play. The subject is the VE-surface that contains the qualia of the observer's consciousness. When the respective hemispheres of two VE-spheres expand *toward* a possible point of encounter, the encounter may or may not occur. It may be pre-empted by the "environment". This happens when either one of the other two hemispheres that are expanding *away* from the point of possible encounter is encountered first, by some other VE-sphere. An encounter with the surface of a VE-sphere at any point causes the collapse of the entire surface. This is in keeping with Tegmark's reasoning that "the object decreases its entropy when it exchanges information with the subject and increases it when it exchanges information with the environment. Loosely speaking, the entropy of the object decreases while you look at it and increases while you don't."(7) To say that "looking at an object" decreases the object's entropy means that an act of observation injects energy into the object, thereby contributing to its continued existence.

Tegmark's paper also mentions the possibility, first suggested by Roger Penrose, that the brain processes associated with consciousness may occur in so-called *microtubules*, which are like hollow cylinders that help brain cells maintain their shape. There are conflicting theories about how nerve impulses might travel through these hollow cylinders. The scenario proposed here suggests that the "cylinders" are the superposed circumferences of the QV-bubbles that exist (and continuously multiply) within the expanding surfaces of VE-spheres. When two of these surfaces encounter each other and collapse, this leads to a *cascade of encounters* between the QV-bubbles in the respective surfaces. The vacuum energy in all of the QV-bubbles is released and rushes into the expanding surface of the new VE-sphere that is created at the point of encounter.

The entire cascade occurs, paradoxically, at a single point - the point of encounter. That point, however, consists of countless points superposed in the dimension of time. The rush of vacuum energy along the line of points occurs - another paradox! - within a "tunnel". It has the diameter of an *oscillating string*. All encounters between VE-surfaces are, as I explained in chapter three, encounters between the thrusts that theoretical physicists have called "strings". They oscillate within all three dimensions of space.

In this scenario, the gluonic light of consciousness created in an atomic nucleus consists of nerve impulses that travel through countless microtubules at the same time. The synchronized nerve impulses occur at no single location in the brain. They are a *field of energy*. Consciousness is produced at countless different points, like a holographic image. When the brain stops functioning at the onset of an NDE, gluonic light is no longer produced in the microtubules. When the last rays of gluonic light travel through the microtubules, the mind experiences itself as traveling through a "*dark tunnel*", as the experience is often described by those who have lived through an NDE.

After watching the efforts to restore life to the clinically dead body for a while, the mind hovering above the body leaves the scene through the dark tunnel and travels toward a bright light that unfolds into a vibrant holographic image. As reported by Stahlhut, when the mind of the architect von Jankovich left the location from where he had been observing the scene of the accident that took his life, he experienced that event as follows: "When I saw that whole scenario, I really was not interested in anything anymore, I wanted to fly away and I was just flying away (...) and I wanted to fly into the center of this light, just like a mosquito flies into the flame of a candle and destroys itself (...) and I was crazy with happiness about the acceleration, faster and faster toward the light (...) and the closer I got, the less it blinded me (...)."**(8)**

The light described by von Jankovich, we can assume, is "pure" gluonic light devoid of any input from actual events occurring outside the clinically dead brain. This final aspect of an NDE is experienced just before the mind leaves its out-of-body location and the brain starts to function normally again. Up until that very last moment of the anomalous experience, the positronium at the location of the out-of-body mind is continuously renewed as a stable configuration. The unobservable energy for this process is supplied, through retrocausation, by the echo waves emitted from the brain awaiting its revival.

The quantum vacuum in the QV-bubbles of VE-surfaces provides the vacuum energy that energizes the renewal of the positronium during an NDE. Since spin, electric charge, and magnetism are created by the vacuum energy in the surfaces of interacting VE-spheres, this explains why the "pseudo-atom" mentioned by Hotson in his description of positronium does not instantly disappear. When a positron and an electron have interacted and are no longer detectable, Hotson pointed out (14/4), "conventional wisdom says that their charges and spins have 'cancelled' and that they have 'annihilated' and are no more". Hotson rejected this conventional wisdom and continued: "(...) since they never get closer to each other than 56,000 times the diameter of a proton, how can they possibly 'cancel and annihilate'? (...) For them to 'annihilate' would be *action at a distance*, a direct violation of causality."**(9)** (Italics by Hotson)

Hotson argued that, while mass can change into energy, and vice versa, mass-energy can neither be created nor destroyed, as stated in the law of conservation, which is one of the fundamental laws of physics. Similarly, he said, the energy of particle spin and of an electric charge cannot simply "disappear". Since the energy of photons, and the effects of gravity and mass, are called *positive energy*, Hotson suggested that when the energy of spin and charge "disappears", it becomes a part of an energy field that consists of *negative energy*. This maintains the basic symmetry of all physical forces. The confusing use of the terms "positive" and "negative", Hotson said, apparently goes back to Benjamin Franklin, "who flipped a coin and chose to designate static electricity with a deficiency of electrons 'positive' and that with a surplus of electrons 'negative'. He assumed correctly (...) that electricity was the (...) flow of some charged entity; but he guessed exactly wrong as to which one. (...) Had he guessed correctly as to *what was flowing*, (...) the electron (...) would have had a *positive* sign."**(10)** (Italics by Hotson.)

When the electrons and positrons interacting in positronium "lose all their positive energy and become undetectable", Hotson argued, they "drop into a negative energy sea", where they "appear to be permanently associated in pairs (...)."**(11)** This is another striking example that the metaphorical images proposed by the Physics of Encounter are in keeping with the concepts used by theoretical physicists. The effects that are "permanently associated in pairs", as interpreted here, are contained in the two hemispheres of an expanding VE-sphere. The surfaces expand into opposite directions of time and contain the undetectable potential energy of electrons and positrons. The surface expanding toward the observer transmits the effect of an *electron*. The surface expanding away from the observer transmits the effect of a *positron*. What Hotson called the sea of "negative energy" contains the unobservable effects of VE-surfaces expanding away from the observer. These surfaces are the *echo waves* of electromagnetic radiation. When they transmit the effect of the positive charge associated with a positron, they transmit the anti-gravity effect of a particle that, as Feynman said, "travels backward in time". Echo waves transmit the effects of backward causation. They are the unobservable source of energy that keeps an out-of-body mind in existence during a Near-Death-Experience.

*NOTES*

(1) Manthey (2006), p. 2
(2) Stahlhut (1993), p. 19
(3) Ibid, p. 17
(4) Hotson (2002), p. 49
(5) Stahlhut (1993), p. 2
(6) Tegmark (1999), pp. 1-15
(7) Ibid, p. 9
(8) Stahlhut (1993), p. 22
(9) Hotson (2002), p. 49
(10) Ibid, pp. 50-52
(11) Ibid, p. 53

# *Chapter 15*

**Motion, changing shapes, visual fields. The stunning maneuvers of UFOs. Balls of light that hover near their observers. Séances, levitation, and the "possessed mind".**

> *If we ask, for instance, whether the position of the electron remains the same, we must say "no";*
> *if we ask whether the electron's position changes with time, we must say "no";*
> *if we ask whether it is in motion, we must say "no".*
> J. Robert Oppenheimer

This chapter will address the puzzle of observed motion that seems to be in stark violation of the known laws of physics. The flight patterns of UFOs are a case in point. The concept of motion is, in itself, a puzzling aspect of the processes that occur at the elementary level of reality. Does anything move inside an atom? The electron shell of an atom contains electrons that make quantum jumps and change their energy when doing so. Physicists have learned to "split the atom", slicing into the events that occur inside an atomic nucleus. This releases nuclear energy. It can be used in an atomic bomb or to produce electric power. Atoms are filled with highly energetic activity. Yet, paradoxically, nothing actually moves while these subatomic events occur.

J. Robert Oppenheimer was the director of the US government's secret laboratory at Los Alamos that developed the atom bomb. The electron to which he referred in the above aphorism is an orbital electron that makes quantum jumps. As explained in chapter two, the words "orbit" and "quantum jump" are just metaphors, two of the many figures of speech used in theoretical physics. Nothing travels in an orbit, nothing jumps. The orbital electron is an energy point. It is a point of encounter between the expanding surfaces of VE-spheres. The wavelength of the electron is the distance between two points of encounter in the electron shell of the atom.

Atoms are the fluctuating vacuum energy in interacting VE-surfaces. They are what Stahlhut (14/2) called "the hardened version of reality". The Physics of Encounter suggests that the anomalous motion of a UFO is like an event inside an atom. Nothing actually moves. To be more precise, the observed event involves atoms interacting through the exchange of photons. Photons, too, are the expanding surfaces of VE-spheres. The characteristics of the surfaces that differentiate photons from matter particles are described in chapter twenty-one.

As explained in chapter thirteen, the events in the nucleus of an atom create the gluonic light of consciousness if they involve an exchange of *bio-photons* between atoms interacting within a physical system that this alive. Before a photon is absorbed into an atom, it is the instantly expanding surface of a VE-sphere. It contains the vacuum energy that exists throughout space. An expanding VE-surface is an undetectable source of energy. UFOs are a physical reality, like the photons exchanged between atoms, and like the atoms that emit and absorb them. UFOs appear on radar screens, create magnetic effects, and leave physical traces at landing sites. Their "flight patterns", however, do not correspond to what is actually happening in physical space. They defy the laws of physics. A UFO may abruptly make a right-angle turn. It may stop instantly, "on a dime", so to speak, or it may hover, motionless, and then suddenly depart, instantly accelerating along a straight line to breathtaking speed.

The physical effects of a UFO are the effects of atoms and of photons, but the "unidentified flying object" is not what it appears to be. It is like a holographic image that has no graspable physical substance. The image is created when the influence of the human mind on physical processes is amplified by the hidden energy of the collective unconscious. I explained the details in chapter eight. The "appearance" of a UFO should be understood in both meanings of the word. When a UFO suddenly appears as a mysterious object in the sky, that anomalous event is triggered by the effect of human consciousness interacting in morphic resonance with the vacuum fluctuations of the collective unconscious. Such objects often look like spaceships because subconscious expectations fuel that conclusion.

Let's look at the elementary level of reality where the quantum vacuum fluctuations of space are, as Puthoff said (7/2), a "blank matrix upon which coherent patterns can be written". During the anomalous mind events associated with the observation of UFOs, the "coherent patterns" are the stunning flight patterns of UFOs and the outward appearance of these objects to observers who are in a state of mind that is like a partial self-hypnosis. UFOs are not what Stahlhut called a "hardened version" of reality. Their uncanny maneuvers are like the special effects of digital animation in state-of-the art movies. The effects create a reality in which, in smooth transitions from one image to another, *everything imaginable* is possible. When the mind, mystified and excited, is strongly focused on a specific event in the sky, shutting out most other observations of normal reality, the imagination kicks in.

Events that create a strong emotional focus are like those that Roger Nelson studied in his Global Consciousness Project. The evidence, he said (8/3), suggests the existence of a "communal shared mind in which we are participants (...)." It is a "consciousness field" in which large numbers of people become "deeply absorbed in one focus". The focus is created by events that "inspire strong coherence of attention and feeling".

As I explained in chapter eight, observing a mysterious object in the sky that is presumed to be a UFO creates this strong focus of attention and feeling. The "communal shared mind in which we are participants" is the collective unconscious. The Physics of Encounter suggests that the patterns of anomalous motion correspond to outlines of the most frequently observed configurations created by man, or by the forces of nature. These are the "coherent patterns" that are most frequently written on what Puthoff called the "blank matrix of quantum vacuum fluctuations". Among the man-made configurations are the straight-line/right-angle structures of buildings, windows, doors, bricks, boxes, sheets of writing paper and many other configurations. The Physics of Encounter suggests that straight lines and right-angles are among the most prominent features of the images stored in the collective unconscious. They are at the "top of the mind" of humans when their subconscious emotions are stirred by exciting or troubling observations associated with recent technological achievements. When inexplicable events are observed in the earth's atmosphere and thought to be spacecraft piloted by alien beings, the collective unconscious kicks in with a powerful dose of imagery that influences what is seen. The collective unconscious shapes the patterns of the observed motion. The straight-line and right-angle maneuvers of UFOs, executed at high speed, are emotionally exciting because they are downright "impossible" and "out of this world".

The presence of UFOs in the atmosphere is often confirmed by the blips' that suddenly appear on radar screens. Radar recordings of the flight patterns, however, are just a chaotic tangle of lines bearing no resemblance whatsoever to the stunning flight patterns of UFOs seen by observers looking upward into the sky. Why the discrepancy? What causes the observed motion of UFOs along the outlines of geometric patterns?

The geometric patterns do not appear on radar screens because they exist only in the mind of the observer, and in the collective unconscious. Other patterns, also stored in the collective unconscious, are characteristic of other phenomena that the currently known laws of physics cannot explain. The mysterious "crop configurations" in farm fields are one example. The explanation suggested here is that they, too, are created by the unobservable effects of the collective unconscious. This unobservable reality is intimately entangled with the observable reality in our universe - the forces of nature.

Among the configurations frequently created by the forces of nature are the crystalline patterns of minerals and the circular effects of whirlwinds and tornados. Storms cannot be the direct cause of what are commonly called crop circles, however, because storms hardly ever occur when and where these circles are created. The other typical crop configurations are shaped like crystals. These are substances in which atoms and molecules are arranged in a regularly repeated definite pattern. A crystal can have the straight-line, right-angle structure of a cube, or have many other shapes consisting of flawless straight lines, among them the rhombic shapes with edges that are parallelograms.

The circular and crystalline patterns appear overnight in the fields of farmers. The event seems to be associated with the previous observation of a UFO, or of ball lightning, in the immediate vicinity. The process that creates these "coherent patterns" presses stems of the crop to the ground without breaking them.

Simeon Hein has suggested an explanation for the circular and crystalline patterns in crop formations. I will skip the complex details, but want to mention that the process involves photons. Hein suggested that sharp angles "create a photonic, crystalline effect that could interact with the light coming into the formation. As this photonically altered light interacts with the plants, it may alter the distribution of electron orbits (...). Secondly, as the electrons' orbital frequency is lowered they become susceptible to spontaneous phase resonance (...)".(1)

As interpreted here, the phase resonance corresponds to the morphic resonance between the fluctuating vacuum energy in the surfaces of photons and the surfaces of other VE-spheres when these VE-surfaces interact and create the reality of particles and atoms. In the Physics of Encounter, morphic resonance is involved not only in the creation of anomalous events, but in all processes that create "hardened" and measurable physical reality from the quantum vacuum fluctuations of space. The interaction between the two levels of reality is a "normal" aspect of the overarching reality we experience as matter, and as our consciousness. The interaction is always an "entanglement" with the undetectable energy of the collective unconscious. It creates an "anomalous" event only when the strength of the morphic resonance exceeds the normal level of statistical probability. There is no sharp dividing line between normal and paranormal events. As I explained, the statistical probability of an anomalous event increases with the size of the VE-sphere that contains the qualia of the observer's consciousness.

The creation of a crop formation and the observation of a UFO are relatively improbable events but are consistent with the laws of nature as interpreted here. Both involve elementary geometric patterns that exist as potential energy in the quantum vacuum fluctuations of space. The patterns exist in the collective unconscious. This scenario is supported by scientists who have investigated Near-Death-Experiences. During an NDE, the human mind is strongly influenced by the effects of the collective unconscious. The out-of-body mind experiences patterns of light and brilliant colors. According to Christian Stahlhut, the psychologist Ronald Siegel pointed out that the abstract images seen during an NDE are composed of elementary geometric structures.(2)

Additional evidence supporting the proposed scenario is provided by reports about the motion of the balls of light commonly assumed to be *ball lightning*. When the luminous sphere is observed moving around in a building (it "poked around", as one witness said), it shows a preference for following the straight lines where vertical and horizontal surfaces meet. Ball lightning is often seen moving along the floor near a wall. The luminous ball also seems to have an affinity toward humans. It often hovers near the person who observes it, or follows that person. It can also move, smoothly, against the wind.

The balls of light, like UFOs, seem to have a biological component. In the scenario proposed by the Physics of Encounter, this means that bio-photons are exchanged between the observer and the observed phenomenon. Here, too, the motion is energized by the mind of the observer and amplified by the potential energy of the collective unconscious. This scenario explains the motion of images in the observer's mind as well as the force that causes the anomalous motion of physical objects. Some examples of the latter are Poltergeist events and the invisible force that causes a table to rise from the floor during a séance (levitation).

Let's look first at the process that creates the observed motion of UFOs. The motion occurs in what neurologists call the visual field. It is created by physical processes involving the nerve cells in the brain. The visual field is like a constantly changing holographic image that exists in the mind. In some cases, the mind may see only a simple geometric pattern, not necessarily all of the details associated with the observation of an actual object. To visualize this scenario, imagine that you are looking (in your mind) at the boundary between two different, uniform colors that fill your entire field of vision, and that the boundary is a straight line. Each of the two colors is an undifferentiated expanse of mental space, and the boundary line is *clearly localized* within that space. In addition to the two colors, the straight line is a prominent feature of what you are seeing.

The straight line is the simplest example of the image you see when you look at the clear-cut boundary of many man-made objects. What happens in your field of vision if the line begins to wiggle like a snake? The motion of images in the mind is always a smooth motion, but that is not the case when the moving image depicts the adventures of Donald Duck, for example, in an animated cartoon created when the movie industry was still in its infancy. There is just no way an artist can draw the outline of images that move like actual objects or living creatures. Every point in one outline would have to be located right next to the point in the outline that exists at the next moment in time, with no space in between.

In a bold analogy, one could say that Donald Duck's jerky antics illustrate the conundrum that physicists have not yet resolved: the incompatibility of relativity theory and quantum mechanics. In the mathematics of relativity, space and time are a *continuum* in which motion is a "smooth" transition from one location to the next. In the theory of quantum mechanics, an object is a *discontinuum*. It consists of elementary particles, and these are "quantized". They are chunks of mass and energy. Most physicists agree that space is not quantized. It does not come in chunks. So what is the smallest unit of space and time if it is not a quantized, measurable unit of "something"?

For the answer, we must look at the process in which mass and gravity effects are created by the particle that Ledermann (10/11) called the "God Particle". As interpreted by the Physics of Encounter, that process is a cascade of encounters between the QV-bubbles superposed in the expanding surfaces of VE-spheres that have encountered each other. The particle created by the final encounter between the increasingly smaller QV-bubbles is the Higgs boson. It is the particle that interested the physicists at the CERN research facilities near Geneva. They built the world's largest particle accelerator to see if it could create that particle in a simulation of the "big bang" that created our universe.

In the scenario proposed by the Physics of Encounter, there is no observable final encounter between the converging surfaces of VE-spheres because each VE-sphere is like a QV-bubble located in the surface of a larger VE-sphere. Every encounter is preceded by a cascade of encounters, and every encounter is only an excerpt of a larger sequence of encounters, like a snapshot of a moment in time. As in every snapshot, the camera lens can capture only a segment of the reality that presents itself at any specific moment. The "camera lens", in this metaphor, is the human mind. It establishes the size, or wavelength, of what Cramer (9/7) called the "inflating bubble" of a photon. There are much bigger "photons", or VE-spheres, but they do not interact with our senses. A VE-sphere that is a photon differs from other VE-spheres because it is created when VE-spheres expand toward a common point of encounter along all four dimensions of space *and of time*. An atom can instantly absorb a photon *in its entirety*. A photon "fits into an atom".

Based on what our senses tell us about reality, physicists have drawn conclusions about the nature of elementary particles. The nature of the Higgs boson cannot be explained by current theories because the *Higgs field* that creates the boson exists between the converging surfaces of VE-spheres, and every convergence *toward* an encounter at any point in space is instantly followed by a divergence of VE-surfaces *away from* the point of encounter. The cascade of encounters never ends. The Higgs boson is part of the quantum vacuum fluctuations of space.

In this scenario, the mass created by a Higgs boson exists at the point in space and time that is an experienced instant in our consciousness. It exists in the hologram that is our visual field. The motion of UFOs within the visual field of a conscious mind occurs at points that are the constantly changing locations of gravity effects. They contain what Ibison (12/9) called a "potential *for* energy". The flight patterns of UFOs in the observer's visual field and the patterns of crop formations are patterns of gravity effects. They are prominently etched into the collective unconscious because they are frequently observed.

A good way to visualize the effect at such gravity points on the motion of UFOs is to imagine that each point is a tiny lightbulb and that each light is turned on separately for a fleeting moment and then turned off again. We see examples of this type of process when we look at the patterns of blinking neon lights used for outdoor advertising. The illusion of motion along the patterned lines is created when the closely spaced lights blink on and off in rapid succession.

The observed effects that seem to be the motion, or "flight patterns" of UFOs are pinpointed patterns of energy that are stored in the quantum fluctuations of the collective unconscious. The patterns involving what seem to be luminous objects in the sky often represent events of strong symbolic and subconscious significance. This includes the *union* of what were separate UFOs at first, the *"birth"* of additional UFOs from a single one, the *disintegration* of UFOs or of the formation in which they were flying, as well as surprising changes of size, shape, and/or color.

The speed of UFOs can vary drastically, depending on the time interval between the "blinking lights" that are the patterns of fluctuating gravity effects. They define the outline and motion of the images in the visual field of our consciousness. The human mind experiences time because the qualia of consciousness contained in the expanding surfaces of VE-spheres are changed when VE-spheres encounter each other. The time interval between the "blinking lights" can be reduced, or increased, depending on the fluctuating size of the interacting VE-spheres.

The metaphorical images in this scenario correspond to the concepts proposed by the physiologist John C. Eccles. After extensive investigations of the relationship between the brain and the mind, for which he received the Nobel prize, Eccles presented what he called "a new philosophy of perception" based on "feature-detection neurones" in the brain. A neuron, as it is usually spelled in the US, is a brain cell. Eccles did not investigate anomalous mind events, only the straightforward question how outside images, impacting on the eye, are transformed by the brain into consciously experienced images. A single "feature-detecting neurone", Eccles argued, can be influenced by "large areas of the visual field", and "this specific response to geometric forms, such as squares, rectangles, triangles, stars, is dependent on the ordered projection onto these feature-detection neurones from complex and hypercomplex neurones sensitive to bright or dark lines or edges of a particular orientation and length and meeting at particular angles".(3)

As interpreted here, the "ordered projection" of features or geometric patterns onto single cells, and the "specific response to geometric forms" at the location of a single brain cell, correspond to the points of light that are turned on and off, one after the other, for brief moments, as described above, creating what appear to be patterns of motion. The points of light correspond to the locations where a neuron "fires", or sends out a measurable biophysical signal. As interpreted here, the entire pattern exists at each location of a "feature-detection neurone", in the same way that a *holographic image* is contained in the points of a holographic plate.

In the scenario proposed by Eccles, and proposed here, the geometric patterns do not exist as three-dimensional configurations in the brain. They are a sequence of signals in the dimension of time. This is in keeping with the reasoning of the philosopher René Descartes, who argued that the manner in which we see is like the method used by a blind man who *feels his way* when he walks with a stick. In the Physics of Encounter, the dimension of time is not like a single straight line. Time is experienced at points of encounter between VE-spheres where the changing qualia of consciousness are created in a sequence of encounters. This corresponds to the process that creates consciously experienced patterns, as Eccles put it, "in a selective and unifying manner". Eccles pointed out that the psychiatrist Jung and the philosopher Popper compared the process to a "searchlight (...). Perhaps a better analogy", he continued, "would be some multiple scanning and probing device that reads out from and selects from the immense and diverse patterns of activity (...in the brain), (...) organizing them into the unity of the conscious experience."**(4)**

As interpreted here, the "scanning and probing device" of the mind, (the metaphorical searchlight), probes or shines into the *collective unconscious*. The "selective and unifying manner" of the process is based on the morphic resonance between the energy of the searchlight and the energy of the events illuminated by it. The selection process corresponds to an encounter between VE-spheres. The morphic resonance at points of encounter creates the qualia of consciousness. The encounters also create the energy of particles. Some of the VE-spheres correspond to the "inflating bubbles" that are the wave aspect of photons, as described by Cramer (9/7). The expanding and collapsing QV-bubbles in the surfaces of VE-spheres are the quantum vacuum fluctuations that pervade our universe.

The collective unconscious is encoded in these fluctuations. In chapter eight, I explained why VE-spheres that originate in the brain may expand into an abnormally large size. Their interaction creates the anomalous events that baffle physicists. The cascade of encounters that occurs at locations where the surfaces of abnormally large VE-spheres interact in morphic resonance releases an abnormally large amount of vacuum energy. This interaction creates the anomalous plasma of ball lightning and of UFOs. I described the process in chapter nine.

The anomalous events that release an abnormally large amount of vacuum energy occur outside the brain. The transformation of this large amount of potential energy into measurable physical energy does not involve the nerve cells of the brain. This scenario of mind/matter interaction proposed by the Physics of Encounter is supported by the "philosophy of perception" proposed by Eccles, as described above and in chapter thirteen. Eccles argued that the mind is an "independent entity" that does not only "read out selectively" from the ongoing activities of the brain "but also modifies these activities" (13/1).

The "independent entity" is the expanding surface of a VE-sphere. The QV-bubbles in the surface energize the expansion and contain the qualia of consciousness. Encounters between the surfaces of VE-spheres that occur within the brain create new qualia and influence the quantum mechanical events in the brain. The encounters are a dialectical process. On the one hand, the vacuum energy in the expanding surfaces of VE-spheres "modifies the ongoing activities" of the brain. The surfaces of VE-spheres contain the potential energy of mind events that have already occurred. This includes the potential energy of the collective unconscious that exists *outside* the brain. On the other hand, the "ongoing activities" *inside* the brain are encounters between the surfaces of VE-spheres that create the quantum mechanical reality of atoms interacting in the nerve cells of the brain. This interaction influences the mind by creating new qualia of consciousness at the points of encounter.

The expanding surfaces of VE-spheres are a twofold reality. They are mind and matter. Surfaces expanding toward the observer contain the potential energy in the wave functions of matter. Surfaces expanding away from the observer contain the potential energy of the observer's mind. This is the energy that can create anomalous events. It is the energy that flows into the collective unconscious.

An important part of the above scenario applies not only to the motion of UFOs that occurs in the mind of the observer but also to the strange "behavior" of the luminous balls that appear on the scene, and remain visible for a while, when ball lightning occurs. In contrast to the geometric flight patterns of UFOs, the motion of the balls of light is *real motion* in physical space. Their observation does not occur in an altered state of consciousness. The balls of light exist as a physical reality at the location where they are observed. Their substance, however, is like that of UFOs. It is the anomalous plasma that is created by the interaction of abnormally large VE-spheres.

Like UFOs, the balls of light are interconnected with the observer through the exchange of bio-photons. The biological component of ball lightning is created when the luminous ball is observed. This explains its affinity to humans. It involves what could be called a *triangulation* of effects. They are the effects of gravity, magnetism, and electric charge. All three effects are created by the interaction between expanding VE-spheres. In the above scenario, they act at right angles to one another, along all three dimensions of space. "Triangulated" effects balance each other out in every direction. When these effects continuously impact on a ball of light, they can prevent it from moving into any direction. The ball of light would hover, motionless, at a certain location in space.

In chapter thirteen, I explained that the exchange of bio-photons, the energy of life, decreases the mass, or weight, of an object. This is what makes the ball of light "float" like a balloon. In the scenario of interacting VE-spheres, gravity and anti-gravity effects are transmitted by the surfaces of VE-spheres expanding into all directions of space and time. The anti-gravity effect is associated with the effect of a positive electric charge. It is transmitted by a VE-surface that expands away from the observer in the dimension of time. A positive charge is associated with a *positron*, the anti-particle that "travels backward in time". A negative charge is associated with the mass of an *electron*. Mass is the impact of a VE-surface that expands toward the observer. A magnetic line of force is created when the VE-surfaces of photons pass through each other, as described in chapter twelve. All three effects, gravity, magnetism, and electric charge oscillate in the electromagnetic waves of light. The oscillations occur simultaneously in all three dimensions of space.

The strange behavior of the balls of light occurs when the normal effects of photons are amplified by the potential energy of the collective unconscious. This is the fluctuating vacuum energy in the surfaces of abnormally large VE-spheres. As explained in chapter nine, the interaction of these undetectable VE-surfaces creates the abnormally large power output of the plasma that is the luminous substance of UFOs and of ball lightning.

The balls of light can *move against the wind* because they consist of atoms that oscillate between being abnormally large and abnormally small. This creates anomalous, ionized plasma. Ions are fragments of atoms, so to speak. The wind blows right through the lattice of ions, like the air that freely moves through screen windows. The air and the wind consist of molecules that are built from atoms, and in that respect are similar to solid matter. The balls of light, on the other hand, can move through solid matter. They can leave a room through the glass of a closed window, for example, or can enter the cabin of an airplane through the closed door of the cockpit. In principle, moving "against" the wind is like moving "through" the wind, or the through the molecules of the air. When the molecular structures are solid matter that physical objects cannot penetrate, the electric charges of the ions in ball lightning briefly change the molecular structure, so that the luminous balls can pass through it.

The balls of light often *hover near human beings* because they share two physical properties with human beings: the left-handed spin and the positive electric charge that is associated with biological matter and with mind events, as explained in chapter thirteen. The shared spin effects create a magnetic attraction. The shared electric charges create the opposite effect. When the two forces are in balance, a ball of light remains in a fixed location relative to a human being. For the same reason, the luminous balls sometimes *follow humans* when they walk around.

The effects of these physical forces symbolize the relationship between biological entities, between the "self" and the "non-self". The awareness of your own identity involves opposite forces. On the one hand, there is the urge to share your thoughts and feelings with someone else (the non-self) who is close to you. "Losing yourself" in the merger of love is one of the impulses of life. On the other hand, there is the impulse to maintain your identity, to maintain the awareness of your own existence as a separate being, to keep the non-self "at a distance".

The interaction between these two forces, the fixation of a location in a highly energized relationship between the "self" and the "non-self" of an object, is evident in the interaction during a spiritualistic session, or séance. The interaction occurs between the person who acts as a "medium", and the object with which the medium "identifies". In Rosemarie Pilkington's description of a séance, the medium, a young girl, levitated a table. It began to float above the floor of the room. An engineer who observed the séance succeeded in pushing the table downward a certain distance, feeling what he called "an elastic resistance", but he could not budge the floating table when he tried to push it toward the girl. "I was much surprised", he reported, "to find that the resistance to push it in that direction was not an elastic one, but of quite a different order. The resistance was a solid or rigid one (...)."**(5)**

This supports the scenario of mind/matter interaction proposed here. The resistance felt by the engineer came from the "medium" who was resisting the merger with what had become, through her anomalous mind event, a paradoxical part of herself: the levitated table. I fully empathize with the skeptics who cannot imagine that the mind of a young girl interacting with the abstract, undetectable reality of the collective unconscious can make a table rise from the floor. But let me point out the relevance of UFO research in this context. UFOs are a product of the collective unconscious. That is the interpretation proposed by the Physics of Encounter. According to the physicist Illobrand von Ludwiger, who regards them as actual flying objects, there have been several reliably reported incidents during which UFOs *briefly lifted automobiles* into the air. A UFO, von Ludwiger argued, is an actual flying object that generates a gravitational field. He surmised that the gravity field of a UFO can briefly pull a car upward, toward itself, in the same way that gravity pulls an object downward when it falls toward the surface of the earth.**(6)**

The Physics of Encounter suggests an interpretation that involves magnetic polarities. By "identifying" with the object that rises into the air, by developing a "resonance or a bond", by "falling in love" with it, as described in a different context by Jahn and Dunne (8/12), a person's mind can influence the motion of an object through psychokinesis (PK). The influence of the mind changes the spin effects in the object to a predominantly left-handed spin and briefly transforms the object into the gravitational equivalent of anomalous biological matter. The influence of the mind, in such cases, is powerfully amplified by the potential energy of the collective unconscious. The overwhelming dominance of left-handed spin and positive electric charges reduces the mass, or weight, of the levitated object. The object is lifted off the floor by the force of repulsion between the changed magnetic polarity of the levitated object and the normal magnetic polarity of our planet. The object briefly acquires the strength of the magnetic polarity associated with the much more encompassing physical system containing the activated energy portion of the collective unconscious.

The process of "identifying" with a levitated object is a dialectical process. In most cases of levitation, the persons who act as a spiritualistic medium do not lose awareness of themselves. They stay "in control" and do not fall into a trance. They *channel* the energy of the collective unconscious into the levitated object, but stoutly maintain their own identity by keeping the levitated object "at a distance", even though it has temporarily become a significant part of their own biological identity and consciousness.

The mind of people who use the energy of the collective unconscious is like the mind of seasoned actresses or actors when they play a role. They must identify themselves with the character they play. They must truly *be* that character. They must *feel* what that "other person" feels, must shed real tears, feel real rage, laugh real laughs, not hollow laughs that are just a facial facade. And yet they must stay in control of what is happening. Their mind allows itself to be energized by the genuine emotions of the character who "comes to life" for the spectators. Allowing the mind to be energized in this way, however, requires energy of a special kind. It is the same kind of energy that comes into play in psychokinesis. As described by Roll (8/13), the mind must make a powerful and conscious effort to "empty itself" of all distractions, allowing itself to be energized by an outside source of energy, and experiencing itself as an integral part of that energy. As interpreted here, that is the collective unconscious. Being "in touch" with the collective unconscious is being in touch with the reality of a mind that exists outside one's own consciousness.

When actresses and actors create a character who comes to life, they influence their own mind to create a reality that did not exist before. The influence of their mind changes the events in the atoms of their brain. In principle, that is psychokinesis: influencing matter through the energy of the mind. The mind of the character they have created influences what they feel and what they do. It makes them laugh and makes them cry. Through the art of acting, the mind "takes possession" of a character, and allows itself to be possessed. A *"possessed mind"*, in the sense used by investigators of anomalous phenomena, may lose control over itself when a disembodied "spirit" takes over. During *poltergeist events*, the mind of the person who unknowingly causes the events through recurrent spontaneous psychokinesis (RSPK) is "possessed" in the sense of being strongly influenced by subconscious emotions. Through morphic resonance, these mind events release some of the enormous energy of the collective unconscious. In *séances*, the person who acts as a "medium" uses an outside source of energy to levitate objects.

All cases in the broad spectrum of anomalous phenomena involve an interaction, or "entanglement", between the two realities in our universe. One reality exists in the space through which we move in our daily lives. The other reality is an undetectable, boundless energy in a space where there is no motion because it contains the timeless reality to which our universe owes its existence. That includes the reality of the conscious human mind.

## NOTES

(1) Hein (2007), p. 14 (Abstracts)
(2) Stahlhut (1993), p. 7
(3) Popper and Eccles (1977), pp. 268 and 271
(4) Ibid., p. 363
(5) Pilkington (2006), p. 131
(6) von Ludwiger (1992), pp. 147 and 149

# Chapter 16

**Scorched earth and cold breezes: the opposite effects of anomalous events. Rapping sounds and distorted voices. "The world is vibration."**

Temperatures and sounds have one thing in common. You feel temperatures and you hear sounds when the molecules of the air move in certain ways. When you hear sounds, the molecules move back and forth, in rhythm with the sound wave that carries the sound through the air to your ears. The molecules transmit information that can be picked up by the eardrum, which starts to vibrate when the sound wave impacts on it. The molecules that do not participate in this kind of motion are like "noise". They just bounce around as they bump into each other and rebound. The heat you feel on your skin when it is exposed to hot air is the equivalent of noise. Hot air contains what physicists call highly "agitated" molecules.

*Heat* can be either the energy of molecular motion or the radiated heat of electromagnetic waves. Radiated heat is pulsating energy that is like the light emitted from atoms, but it may have shorter or longer wavelengths that are not perceivable as light. *Sounds* and radiation are closely related. Since the mind can influence the motion of objects and of elementary particles through *psychokinesis* (consciously or subconsciously), we need to look at the question whether anomalous mind events can also produce sounds and cause abnormal temperatures. The answer is yes. Participants in séances have reported that "spirits" are able to produce rapping sounds. During séances, musical sounds and strange voices are often heard. As to temperatures, scorched earth that was apparently exposed to extreme heat has been found at locations where UFOs reportedly landed. Paradoxically, the opposite kind of temperature effects are often reported by people who saw a UFO at close range and said that they encountered "aliens". They had the uncomfortable sensation of a sudden drop in temperature. Similar sensations of cold air have been reported by participants in séances during which a table tilts or rises into the air. In a case described by Rosemarie Pilkington, the participants "felt cold breezes flowing upward from the center of the tabletop just before tiltings took place".(1)

David J. Turner described an incident involving ball lightning that occurred at the end of the 19th century. A group of seamen on a sailing ship had observed that "a large and very powerful lightning-ball (...) had fallen into the sea alongside the vessel and caused huge waves to break over it". Standing on deck they "felt a stifling heat", and at the same time they saw that the rigging for the sails at the top of the masts had frozen. Turner suggested that the sailors might have been located in a specific zone of the ball of light, while the freezing occurred in a different zone.(2)

This chapter will examine the above examples of anomalous temperatures and sounds, and several others, in detail. Of particular interest is the paradox that opposite effects are often created by the same type of anomalous event. In the scenario of mind/matter interaction proposed here, the observed effects depend on the location of the observer. Anomalous events are created at points of encounter between abnormally large VE-spheres. The observed effects created by this interaction differ, depending on whether the observer is located at the boundary of the VE-sphere or inside the VE-sphere.

An observed effect is information about physical reality. It can be transmitted by the two types of wave that I mentioned at the beginning of this chapter. The more recent technological developments utilize what could be called a "yes or no" flow of energy at a specific point. This is the digital principle used for transmitting images and sounds. There are only two digits, the number "one" and the number "zero". This corresponds to the principle of oscillating strings. The endpoint of a string is like a metaphorical point of light. It is either "on" or "off", either "one" or "zero". When it is "on" (energized), it turns itself "off" by providing energy for the other endpoint. The energy oscillates between the two endpoints.

The "yes or no" flow of energy corresponds to the random *binary process* described by Moddel and Walsh (12/1) to explain the results of their experiment, which showed that the mind can influence the transparency of glass. At any given point on the surface of the glass, a photon impacting on the surface is in a yes-or-no situation. It either passes through the glass or is reflected. For transparent glass, the average percentage of transmission versus reflection is 92% to 8%. Through psychokinesis, the mind can change that ratio.

The interaction between two VE-spheres at a point of encounter is a binary process. At that point, a new VE-sphere is either created or it is not created. The event occurs only if the two surfaces encounter each other in morphic resonance. The resonance involves the vibrational energy of waves. The expanding and collapsing surfaces of VE-spheres are waves. The surfaces are energized by expanding and collapsing QV-bubbles. Small VE-spheres are like bubbles in the surfaces of larger VE-spheres. There is no limit to the size of VE-spheres. Waves are superposed on waves.

The interaction between VE-spheres can transform radiated heat into the heat of molecular motion. This often occurs on the inside of automobiles parked in the sun, with closed windows. The radiated heat of the sun passes through the glass windows together with the sunlight and is absorbed into the molecular substance of the automobile seats, the dashboard, and the steering wheel. They become unbearably hot to the touch. The heat is trapped as molecular motion inside the car, and the temperature of the air inside the car rises to an uncomfortable level.

Let's look first at the paradox that encounters between VE-spheres either create heat or reduce temperatures. The encounters are interactions between mind and matter because the QV-bubbles in the surfaces of VE-spheres contain the qualia of consciousness. As described in chapter eight, anomalous mind events involve abnormally large VE-spheres. Chapter nine described how encounters between such large VE-spheres produce the abnormally high power output of the light radiated by UFOs and ball lightning.

The Physics of Encounter suggests that encounters between QV-bubbles, expanding and collapsing at points of encounter in the surfaces of VE-spheres, create "temperatures" in much the same way that the random encounters between air molecules determine the measurable temperature of the air. In the air, the temperature is a measurable physical reality. The fluctuating vacuum energy in the quantum vacuum of QV-bubbles, on the other hand, is a non-detectable temperature. It is potential heat energy that is like the radiated heat of waves. The waves are the expanding surfaces of VE-spheres.

The experiments of Bernard Haisch and Alfonso Rueda support the above scenario. Describing what they called a "quantum vacuum basis for gravitation", they reported that an "accelerating detector experiences a pseudo-thermal radiation spectrum whose 'temperature' is proportional to the acceleration".(3) The "pseudo-thermal radiation", as interpreted here, is the potential heat energy mentioned above. The "acceleration" is the creation of a higher frequency of encounters between QV-bubbles. This is the same process that occurs when a spaceship accelerates as it travels through space. The acceleration increases the number of encounters between the QV-bubbles in the vacuum fluctuations of space and the QV-bubbles in the surfaces of the interacting VE-spheres that are the substance of the spaceship.

The higher number of encounters prevents the VE-spheres from expanding into the size they would otherwise attain. VE-spheres, and the QV-bubbles in their surfaces, collapse when they encounter each other. The high frequency of encounters decreases the size of the VE-spheres. As with gas molecules, the frequency of the encounters between QV-bubbles depends on the number of QV-bubbles in a given volume of space. The frequency of encounters between many small bubbles is higher than the frequency of encounters between just a few large bubbles. The smaller the bubbles, the higher the potential heat energy. It is not a measurable temperature, but the interactions between the bubbles become real temperatures when the interactions between VE-surfaces create particles. Interacting particles can become atoms, and interacting atoms can become molecules.

The potential heat energy exists at the boundary of VE-spheres. That is where an encounter between expanding VE-spheres creates a cascade of encounters between the QV-bubbles in the expanding surfaces of the VE-spheres. The interaction between the expanding and collapsing QV-bubbles is potential heat energy. If the VE-spheres are abnormally large, the potential heat energy at the boundary exceeds the normal equilibrium of interacting QV-bubbles that are the quantum vacuum fluctuations of space. In such cases, the cascade of encounters between the QV-bubbles in the surfaces of the VE-spheres involves an abnormally high number of encounters between QV-bubbles, and these QV-bubbles are abnormally small.

This scenario explains why the earth is scorched by intense heat at a location that appears to be the landing site of a UFO. The heat is created at that location because that is where the surface, or boundary, of an abnormally large VE-sphere interacts not only with the surface of another VE-sphere of equal size, but also with the surface of the earth. The *potential* heat effect of that interaction becomes an *actual* heat effect when the QV-bubbles interact with the physical reality of the atoms that are the substance of the earth.

Because of the *binary* ("yes/no") nature of interactions between VE-surfaces, the interaction with the QV-bubbles in the substance of the earth increases the *probability* of encounters that influence the atoms and molecules of the earth and create heat. The molecules in the substance of the earth are densely packed, whereas the molecules of the air are much more loosely spaced.

In this scenario, the potential energy of the observer's mind, amplified by the vacuum energy of the collective unconscious, is an integral part of the physical reality of a UFO. The *boundary* of the abnormally large VE-sphere is continuously re-created while the anomalous mind events of the observer are energized by the collective unconscious. The location where the UFO creates an *effect* (scorched earth) and the location where this effect *originates* (in the brain of the observer) are two different locations. In contrast to normal physical reality, no detectable signal connects the two locations. The process is the same that occurs when staring at a person makes that person aware of being stared at. The effect creates an event in the brain of the person being stared at. The origin of the effect is the brain of the person who is doing the staring. The two locations are connected in morphic resonance by the instantly expanding surfaces of VE-spheres.

Now let's look at the reason why the observers of a UFO often experience the opposite of a heat effect. Seeing a UFO or encountering "aliens" at close range is often associated with the uncomfortable sensation of a drop in temperature. The air suddenly seems to become cold. The Physics of Encounter suggests that the drop in temperature results from a process that is, in principle, like the process in *refrigerators using the compression system*. In such refrigerators, gas is compressed and the temperature of the compressed gas rises. The gas then passes to a condenser, where heat is carried away by water or air passing over a coil. The principle used in this process is *evaporation*. If you have traveled in an airplane and used the moisturized towel that a flight attendant gave you, you experienced the effects of evaporation. The coolness you felt when you placed the towel on your skin was caused by the evaporation of the moisture in the towel. The moisture in such towels is a liquid that rapidly changes into vapor. Examples of vapor are fog, mist, or steam. Vapor consists of moisture particles floating in the air.

The normal body heat stored in your skin produces a cooling effect because it causes the evaporation that carries off heat. The molecules that leave the evaporating liquid in a moisturized towel carry off heat into the surrounding space. The heat dissipates, or is scattered. The compression of gas that starts the cooling process in a refrigerator is like the "compression" of QV-bubbles into a smaller size. The compression is caused by anomalous mind events that create abnormally large VE-spheres. The compression occurs in the space between the surfaces of VE-surfaces that converge on a point of encounter. The space is "compressed" because it is the space between surfaces that approach each other. The cascade of encounters in that space creates abnormally small QV-bubbles.

During the observation of a UFO, or any other anomalous event, the potential energy of the collective unconscious influences events at *two locations*: at the *boundary* of the abnormally large VE sphere associated with the observer's mind event, and at the *origin* of that sphere, which is continuously re-created and energized by the collective unconscious. The events at the two locations occur in two different types of space. They are what I have called *VE-space* and *QM-space*. VE-space contains the vacuum energy of interacting VE-surfaces, of the human mind, and the collective unconscious. QM-space contains the quantum mechanical reality of interacting particles and atoms. The mind of a person who observes a UFO exists in VE-space. The scorching of the earth at the landing site of a UFO occurs in QM-space. The same holds true for the cooling effect that occurs when the "compressed" QV-bubbles in the mind of the observer "evaporate". The cooling effect occurs in QM-space.

In both cases, events in the VE-space of the mind and in the QM-space of detectable physical reality become entangled. The QV-bubbles that are compressed in the mind of the person who observes a UFO then "evaporate" into the QM-space that contains the molecules of the air in which the body of the observer is located. The metaphorical evaporation corresponds to the dissipation or scattering of heat energy in the air that passes over the compressed gas in a refrigerator. The abnormally small ("compressed") QV-bubbles in the mind of the observer expand into the space occupied by QV-bubbles of normal size. This introduces additional vacuum energy into the space between the molecules of the air and the molecules of dust, vapor, and other bits of matter floating in the air. In the vacuum fluctuations of space, encounters between QV-bubbles create new-QV bubbles. Small bubbles expand until they encounter bubbles of normal size in morphic resonance. This re-establishes the equilibrium in the space surrounding the observer's body.

The expansion of the abnormally small QV-bubbles into the air surrounding the observer of a UFO increases the separation between the molecules of the vapor in the air. The normal moisture in the observer's skin evaporates. The molecules in the air are pushed apart. Real heat is dissipated and the observers of UFOs feel the cooling effect on their skin. The effect is experienced only if the size of the QV-bubbles is significantly reduced. This is the case if the emotional impact created by the observation of a UFO is very strong.

In the proposed scenario, an anomalous temperature effect either marks the end of the UFO's existence as a "flying object" or corresponds to a shift in the focus of attention. The attention may wander off temporarily when the view of the UFO that has "landed" is obstructed, usually by trees, a hill, or a building. The heat effect occurs at that time and at that location, when the observer is unable to see the UFO or is not looking. A radical shift of attention, accompanied by strong emotions, occurs when the UFO experience involves an encounter with an "alien being".

Since the temperature effect occurs only when the observer is not looking, the above scenario explains why the plasma of ball lightning is *not hot* while it is being observed. It may hover near a human being, or move along the aisle of an airplane after passing through the door of the cockpit without leaving a trace. The scenario also explains why UFOs that are observed while they are in flight do not give off any heat and only illuminate the area, even when they are flying at a low altitude.

The Physics of Encounter suggests that the heat of a UFO that scorches the earth, or blackens the bark of trees at a landing site, also affects the air that surrounds a UFO at the moment when the existence of the UFO ends. That type of event may occur in mid-air, while the UFO has no contact with solid matter. This explains the occasional reports that black, soot-like particles had apparently rained down from the sky where UFOs were observed. As interpreted here, such material is produced when the hot radiation of a UFO burns the dust particles and other bits of matter that are suspended in the air. The heat creates violent interactions between the molecules in polluted air, blending them into the coarse powder of a charred substance.

This scenario also explains the enamel-like coating of iron oxide that scientists, as reported by Martin Zips, have detected on the flattened stems of crop formations. This substance is created only at extremely high temperatures.(4) The Physics of Encounter suggests that a UFO contributes to the creation of a crop formation when the UFO ceases to exist. Crop formations are created when there are no acts of observation, for example in the middle of the night, or at the very end of an observed event involving the energy of an anomalous light, as the following example shows.

An event involving an anomalous light that disappeared while it was being observed occurred in Uruguay in 1977. As reported in *Phänomene,* a farmer who was out in the field looking after his cattle at four o'clock in the morning noticed that his power generator suddenly stopped working. Then he saw that a swirling, orange-colored light was descending on his field. The farmer's dog ran toward the light and began howling in pain. The farmer felt heat and stood paralyzed. As soon as the light had touched the ground, creating what the farmer described as white flashes of lightning, it disappeared and the generator started to work again. The farmhands and the farmer ran to the location where they had seen the light and saw a classical crop circle, about 40 feet in diameter. The dog was lying there, dead. An autopsy showed that its inner organs had been destroyed by extreme heat of the kind created in a microwave.(5)

Now let's turn to the strange events witnessed by the seamen on the 19th century sailing ship, described at the beginning of this chapter. These events, too, are evidence that the effects involving an anomalous ball of light occur at *two locations* at the same time. As explained above, one location is in VE-space, where the ball of light is continuously re-created, and the other location is in QM-space, where the heat effects of the plasma occur when the ball of light ceases to exist.

During the events observed by the seamen, one location was the deck of the ship where the seamen stood when they felt abnormal heat after a ball of light had fallen into the sea near their ship. The other location was at the top of the masts where the rigging had frozen. The radiated heat of the ball of light, causing huge and violent waves where it fell into the water, was transmitted as molecular heat to the deck of the ship by the molecules of the air. The frozen riggings at the top of the masts were at the location where the ball of light was re-created, one final time, while moving through the air. The drop in temperature at that location was created by the process described above.

Here is another example of the heat effect that occurs when a UFO is still moving through the air and produces the soot-like black powder that I have already described. In 1988, an Australian family from Perth driving along a desert road at night observed a UFO that moved toward them. As reported by *Phänomene*, the occupants of the car were Fay Knowles and her three sons between the ages of 18 and 24. Her son Sean was driving. When he noticed lights above the ground ahead of them, he thought they were automobile headlights. He began to wonder, however, when he saw that the lights seemed to be "hopping around". Then they all saw that the lights had combined into a single source of light that was coming straight at them. Sometimes it was in front of the car, and sometimes behind the car. Suddenly, it was no longer visible. Fearful, and thinking that the light might be hovering above the car, Fay Knowles opened a window and put her hand on the roof of the car. There, she felt that she was touching something soft and rubbery. She later noticed that her hand was covered with what appeared to be soot.(6)

There was more to the events caused by the strange light in the Australian desert, however. After Fay Knowles had stuck her hand out of the car window, the car began to shake violently. It seemed to be "sucked up into the air and then spit out again", Fay said. When the car dropped back to the ground, a tire burst. There was a humming noise outside. All four occupants of the car noticed that their voices were oddly distorted. They ran to some nearby bushes to hide. The light had moved away and had split into several separate lights, which then dissolved. After changing the tire, the Knowles family hurriedly drove on. On the way, they stopped at a police station and described the incident. The police informed UFORA (UFO Research Australia). Their investigators interviewed several witnesses who had also seen lights that behaved oddly and who had experienced what they described as abnormal weather while the nightly sky was calm and clear. The investigators confirmed that a substance that looked like dust, or ashes, was found inside the car of the Knowles family and that there were dents in the roof of the car.(7)

The various aspects of these events can be explained by the scenario proposed by the Physics of Encounter. I will describe the anomalous vibrations that created the humming sound and distorted the voices of the Knowles family later in this chapter, as well as the process that briefly lifted the car into the air and dropped it again. It is an oscillating gravity effect, amplified by the energy of the collective unconscious. Abnormal weather effects, like a sudden wind in calm air, can occur when the radiated energy of a UFO heats the air. Hot air rises and is replaced by cooler air that rushes in to fill the partial void. The rush of air is wind. Distortions of observed images through the effects of heat occur when an object and its observer are separated by the turbulence of hot air. Such effects are observed by drivers looking at the hot air rising from the road in front of them while traveling on an asphalt surface that is heated by intense sunlight. This is why the lights observed by the Knowles family were "hopping around".

The cold breeze rising from a table tilted during a séance, as described by Pilkington (16/1), can also be explained by the scenario of VE-spheres interacting at two separate but interconnected locations. One location is the location of the person who is acting as the "medium" in the séance and "channels" the energy of the collective unconscious into the atoms of the tilting table. The other location is the boundary of the VE-spheres that are created by the anomalous mind events of the "medium". This is the location of the tilting table. The instantly expanding surfaces of these VE-spheres are bio-photons. The vacuum energy in the VE-surfaces is undetectable because bio-photons are created by the mind events of the observer and emitted from the atoms of the observer's brain. They are directly absorbed by the atoms of the observed object. No other atoms are involved.

The scenario of atoms interacting in this way does not only provide an explanation for the creation of anomalous *temperatures*, but also for the creation of anomalous *sounds*. The expanding surfaces of VE-spheres contain the potential energy for the creation of all detectable physical events. Let me start with the fluctuation of undetectable gravity effects associated with UFOs. They can influence very heavy objects or create waves in an ocean. When objects are lifted and fall down again, or when waves of water are churning, we hear sounds. Sounds are created when matter impacts on matter. Those are normal sounds. Anomalous sounds occur when no detectable matter is involved.

The UFO that "sucked up" a car in Australian desert and then "spit it out again" created not only a crashing sound. That was a normal sound created by an anomalous event. The UFO also created a humming noise and distorted the voices of the family in the car. Those were anomalous sound events. In a fact-studded book summarizing the state of UFO research, Illobrand von Ludwiger pointed out that in several reported incidents, UFOs had tipped over automobiles or briefly lifted them. In 1958, on the Atlantic coast of Brazil, several fishers saw a round and flat object that descended from the sky and then hovered above the surface of the ocean. It had the diameter "of a circus tent". Underneath the object, the water seemed to be pulled upwards. It bubbled and looked as though it was boiling. A hill of water had formed underneath the object.(8)

According to von Ludwiger, many physicists assume that a UFO generates its own gravitational field, which counteracts the gravitational attraction of the earth. The Physics of Encounter suggests that the gravitational field generated by a UFO *fluctuates*. The location of the anomalous gravitational field fluctuates between the origin and the boundary of an abnormally large VE-sphere. Because of these fluctuations, an automobile may be lifted into the air in one instant, and then dropped again in the next instant. This happens only if the automobile is located precisely at the boundary of the VE-sphere. The fluctuation explains the paradoxical events mentioned by von Ludwiger, during which some observers of a UFO are briefly lifted into the air, whereas others reported that they were pressed to the ground. The branches of trees, too, are occasionally pushed downward, or move back and forth, as if a strong wind were blowing through the branches.(9) The effect of being pinned to the ground by the light of a UFO was also mentioned by the Russian pilot Vyatkin (11/6), who reported that it happened to the police chief of Voronezh.

The fluctuation of anomalous gravity effects that causes the up-and-down movement of briefly lifted objects and the back-and-forth movement of branches can also make the leaves on those branches rustle, or cause any other event that produces sound waves when matter impacts on matter. Sound waves are the rhythmic motion of air molecules. When fluctuating gravity effects interfere with the differentiated sound waves of voices, the voices are distorted, as described in the report (16/7) about the Knowles family who observed a UFO while driving through an Australian desert. The fluctuations can produce sound waves by causing corresponding vibrations in the air. The correspondence is established in the relatively rare cases when there is a morphic resonance between the fluctuating vacuum energy in VE-space and the quantum mechanical events in QM-space. The quantum mechanical events involve the interaction of particles in the atoms and molecules of physical matter. The humming noise heard by the Knowles family is an example of such morphic resonance. It is an "entanglement" of events in VE-space and QM-space.

In a comprehensive article on humming sounds, which often occur during events involving UFOs, Kurt Diedrich addressed the question whether *events outside the brain* can create the sensation of such sounds without the presence of sound waves. He showed that the proposed theoretical scenarios for this type of event are not in accord with the known laws of physics. Diedrich's argument: sensations of sound can occur without the presence of sound waves only if they are created *inside* the brain or in the sensory apparatus that transmits the corresponding nerve impulses from the ear to the brain.(10) An example of this is an affliction called tinnitus. People who have this affliction experience what seems like a ringing or buzzing in the ear without an external cause. Experiments have shown that direct stimulation of the brain cells can create a variety of sensations, including the sensation of sounds. But electromagnetic fields that originate outside the brain and are not transformed into sound waves cannot create the sensation of hearing a sound.

To be experienced as sound, vibrations of the air have to make the eardrum vibrate, which in turn transforms the vibrating energy into nerve impulses. The eardrum is extremely sensitive to vibrations of the air. According to Diedrich, it can respond to vibrations across extremely short distances corresponding to the diameter of a hydrogen atom. If, through some unknown mechanism, the energy fluctuations of electromagnetic waves were transformed into the rhythmic motions of air molecules, the vibrating energy of that motion would be even lower than that and could not be detected by the ear.

Diedrich pointed out that some people who said that they encountered UFOs at close range, which sometimes included encounters with alien beings, have claimed that they heard voices in their head. They described the voices as actual sounds, not as messages or thoughts transmitted by telepathy, without words and sounds, as is usually reported by people who say they experienced encounters with aliens. In his concluding remarks, Diedrich suggested that the transmission of sounds without sound waves, and the transmission of information by telepathy, occurs in an "invisible energy field" that projects information directly into the brain. That field, he said, could be "a field in which consciousness itself is rooted".(11)

Diedrich's suggestion expresses the essence of the scenario proposed by the Physics of Encounter. The "invisible energy field" is the vacuum energy in the surfaces of VE-spheres. The surfaces of VE-spheres contain QV-bubbles that energize the expansion of the surfaces. The oscillation of the strings in the quantum vacuum of a QV-bubble create the basic constituents of consciousness (qualia). Encounters between the surfaces of VE-spheres, therefore, are *encounters between mind and matter*.

The encounters occur in the Higgs field between the converging surfaces and generates the mass of particles, as described by Lederman (10/6). The wave patterns in the interacting surfaces are vibrations in an undetectable reality. All observable reality in our universe is created by these vibrations. This scenario is supported by the theory of physical reality presented by David Bohm and Karl Pribram. As summarized by Stahlhut, their theory is based on the evidence that "the world is vibration. For one hundred years, this is what modern physics has been trying to instill in our minds, mostly in vain. (...) There is no set of building blocks (...). Elementary particles are (...) elementary waves (...). The world of quantum physics, of uncertainty and haziness, of jumps and potentiality, is a world in which reality literally dissolves as we try to grasp it. That world is hard to fathom. The British physicist David Bohm suggests an even more radical revision of our thing-thinking. He says that the ultimate reality does not consist of measurements made in a laboratory, of more or less stable quantum phenomena. It is better described by concepts such as an ocean of energy (...). Behind the observable order of things, (...) there is a (...) reality that is independent of space and time. It consists of a constantly flowing motion that continuously re-creates the reality we experience at the present moment."(12)

Karl Pribram, a neurophysiologist, developed a holographic theory for the events in a human brain. According to Stahlhut, Pribram suggests that the brain responds to complex energy patterns of interacting waves. "Therefore, the brain recording of an event that occurs at any one moment is like the plate containing a holographic image. Its special characteristic: every single point on that plate contains information about the entire image." In the combined scenarios of Pribram and Bohm, as summarized by Stahlhut, "the brain creates (...) 'hardened' reality. It interprets wave frequencies originating in a reality that lies beyond space and time."(13)

By creating "hardened reality" from undetectable vibrational energy, the human mind can produce the physical reality of sound waves and temperatures when it interacts with the vibrational energy of the collective unconscious. The creation of rapping sounds during a séance, for example, occurs in much the same way as the creation of a UFO. The mind triggers the observed effect that it experiences like a holographic image. But the mass of a UFO, as an atmospheric event outside the mind, is "real". It has an effect on radar screens. The magnetic effect of a UFO and the abnormally large power output of the light it radiates also attest to its physical reality. Similarly, the holographic image of a "ghost" that produces rapping sounds on a table "hardens" the reality of the ghost. The source of the rapping sounds is an anomalous mass, a configuration of gravity effects, that can impact on the surface of a table. It is no more a ghost than a UFO is an extraterrestrial spaceship.

The vibrational energy of a "reality that lies beyond space and time", as Pribram described it, also provides a key to understanding *Near-Death-Experiences* (NDEs). Christian Stahlhut pointed out that many people who were brought back to life after their clinical death, caused by an accident or a heart attack, experienced their minds as floating outside their body, immersed in what they described as vibrating energy. They heard harmonious music, saw brilliant colors, and were attracted toward a vibrant God-like light. The scenario proposed by Kenneth Ring, as summarized by Stahlhut, is that NDEs are "excursions into a realm outside of space and time, filled with energy patterns of higher frequency, which our mind transforms into vivid images". To describe his scenario, Ring, according to Stahlhut, also used the example of a holographic image: "I believe this is a realm that is produced by interacting thought structures. These structures (...) combine into patterns in the same way that patterns of wave interference create images on a holographic plate. And in the same way that a holographic image appears totally real when a laser beam illuminates it, the images produced by interacting thought patterns appear to be real."**(14)**

The question posed by this NDE scenario has religious implications. If an NDE provides information that exists beyond space and time, what kind of reality are the encounters, during an NDE, with loved ones who have died? Stahlhut put the question this way: "Are the loved ones who have died, but who are encountered during an NDE, only stored patterns of wave interference or are they independently existing beings? Are they both?"**(15)**

Stahlhut's question cannot be answered by scientific experiments. But the facts he presented in his radio broadcast about NDEs support the scenario proposed here. The out-of-body mind, experienced by people as their own mind while they are clinically dead, and the person who is encountered as a holographic image during an NDE, are both realities in the collective unconscious. The Physics of Encounter suggests that the encounters during an NDE are examples of what physicists call *nonlocal proximity*. The two minds encountering each other during an NDE are close to each other in the sense that they share certain "vibes", or vibrational patterns. This corresponds to the *morphic resonance* described by Sheldrake.

If there are bodiless minds in a realm that exists beyond space and time, it is not a far-fetched notion that morphic resonance could enable such minds to provide information to the living through the same process that occurs during telepathy, or during séances. The open question is whether such information is *transmitted by a source* that is an independently existing consciousness.

This question was addressed by Rosemarie Pilkington in an appendix to her book, in which she explained why she did not include the research "being done on Electronic Voice Phenomena (EVP) and Instrumental Transcommunication (ITC), electronic voices and images purported to be imprinted on tape or seen/heard on radios, telephones, computers and television sets. Whether these sounds and images are proof of survival after death or psychokinetic effects created by the living is still a subject of controversy."(16)

One of the believers in such phenomena was the Thomas A. Edison. He invented the electric light bulb and enabled New York City to install the first electric street lights in 1882. He is probably best known for his invention of a sound recording device called a phonograph. In an interview published by the *Scientific American* in 1920, according to *Phänomene*, Edison proposed the development of a device for communicating with the dead and actually started working on it. "If our personality survives death," he said, "it is only logical and scientific to assume that memory and intellect as well as other abilities and the knowledge that we acquire during our lifetime remain intact. (...) It is reasonable to conclude, therefore, that those who have left this earth will desire to contact those whom they have left behind. I am inclined to believe that our personality can influence matter from the beyond."(17)

Edison was unable to build the device he had envisaged, but his thinking shows that the human mind contains a deeply rooted desire to fathom the reality from which it arises. Edison's interview with a prestigious publication also illustrates the need to distinguish between assumptions that may be "reasonable" but are not necessarily "scientific".

Many anomalous phenomena cannot be confirmed by experiment. They occur unpredictably, based on the morphic resonance that creates an entanglement of two realities. One of these is the "hardened" physical reality that can be studied by the tried and proven methods of scientific investigation. The other reality is undetectable, but it is reasonable to assume that it exists. We need to examine the plausibility of reports about anomalous events that strengthen reasonable assumptions. In short, we need to develop a more flexible and comprehensive approach if we want to deepen our understanding of mind/matter interactions.

*I simply believe that some part of the human soul or self is not subject to the laws of space and time.*
Carl Gustav Jung

232

<u>NOTES</u>
(1) Pilkington (2006), p. 220
(2) Turner (2003), p. 462
(3) Haisch and Rueda (2001), p. 12 (Abstracts)
(4) Zips (2005), p. 17
(5) *Phänomene* (1994), p. 59
(6) Ibid, p. 20
(7) Ibid, p. 20
(8) von Ludwiger (1992), pp. 147 and 148
(9) Ibid., pp. 147 and 148
(10) Diedrich (2006), pp. 7-11
(11) Ibid, pp. 10 and 11
(12) Stahlhut (1993), pp. 15-16
(13) Ibid., pp. 16-17
(14) Ibid., pp. 17-18
(15) Ibid., p. 18
(16) Pilkington (2006), p. 236
(17) *Phänomene* (1994), p. 38

# Chapter 17

**Time warps and gravitational fields. UFOs, "frozen sounds" and prematurely aged plants at landing sites. Near-Death-Experiences, iPods and accelerating spaceships.**

*All matter is just a mass of stable light.*
Sri Aurobindo

According to current wisdom, time is a dimension. It is one of the four dimensions in which matter exists. Matter has a more or less clearly defined *expanse* in space. It consists of elementary building blocks (particles) that exist for a certain *length* of time. But, as the physicist Avshalom Elitzur observed (3/1), "there is more to time than just a dimension". We can see the dimensions of matter in space and time at a glance because the atoms of matter emit light. Time elapses as we observe the events that are created by the quantum mechanical interactions between the building blocks of matter.

The Physics of Encounter suggests that time, too, consists of what could metaphorically be called building blocks. They are building blocks without physical "substance". Time and the elementary reality of matter are both hard to grasp, in the figurative and in the literal sense. The particles that are the building blocks of *matter* are measurable portions of energy. How do physicists measure *time*? Time elapses as matter decays. Large and heavy particles disintegrate into their smaller constituents. Physicists are able to see what their measuring instruments tell them about these minuscule energy changes because atoms emit photons, the particles of light.

The building blocks of time are contained in photons. Photons are the vacuum energy in the expanding surfaces of interacting VE-spheres. The QV-bubbles in these surfaces contain the elementary constituents of consciousness called *qualia*. Successive moments of consciousness are *differentiated qualia* created at points of encounter between the surfaces of VE-spheres. A moment of consciousness that consists of a large number of *identical qualia* would just be a "longer moment" that delays the beginning of the next moment of experienced time.

The flow of experienced time influences the *intensity* of the experience. While you are enjoying yourself, "time flies". When you are bored, or "clock-watching" at the office, waiting for the appropriate time to leave, time is stubbornly sluggish. The hands of the clock seem to move very slowly.

The number of qualia, and the intensity with which they are experienced, are interrelated aspects of the same process.

What creates a new and different mix of qualia? What determines the intensity with which qualia are experienced? The Physics of Encounter suggests that a new and different mix is created in the nucleus of atoms interacting through the exchange of photons. The mix depends on the number and the size of the QV-bubbles participating in the cascade of encounters that occurs when the surfaces of VE-spheres interact. The vacuum energy in the QV-bubbles is absorbed into the atom. The intensity of the experience is determined by the number of encounters that create the same mix of qualia at any one moment. When that number is abnormally large, the experience at that moment is intense because the effect of each encounter impacts on the mind. If the abnormally large number of impacts create abnormally long moments consisting of unchanged qualia, this heightens the effect of each experienced moment. During anomalous mind events, there are *fewer* differentiated moments, but they are more intense.

The experience of encountering "aliens" is intense because it is focused on a small number of differentiated moments. The people who experience such encounters are not aware that the flow of time in their mind has slowed, relative to the reality outside their mind. They are captives of their own experience. They have no outside point of reference. They are like travelers in a high-speed spaceship who are in a "time warp". When they look at the clock in their spaceship, the hands of the clock move normally.

Nevertheless, the effects that change the flow of experienced time provide some clues about their cause. Time is relative, as Einstein said, not only when you are traveling in a spaceship. The intensity of an anomalous experience is a very clear indication that something extraordinary is happening. When time is slowed through the process described above, the experience is "much more real than ordinary reality". This apt remark was made by someone who said he had been abducted by aliens. The descriptions of Near-Death-Experiences are strikingly similar. An NDE, too, involves what could be called "compressed time". The entire Near-Death-Experience is reflected back in time from an out-of-body location of the mind into the brain that was clinically dead and has started to function again. I explained the process in chapter fourteen.

Another example of a time warp is the so-called "solid light" that is sometimes associated with the observation of a UFO. The experience was described by the Russian pilot Vyatkin (11/6). The beam of light that, as he said, was "splintered" by the wing of his plane was like *frozen light*. The pilot was in a partial trance. The light he saw when his plane flew toward it was like a freeze-frame image of a film. When his plane flew through the light, what he saw was like a film that was not projected at normal speed. The splintering of the light occurred in slow motion.

An encounter with a UFO may also involve what could be called *frozen sound*. Suddenly, there is an eerie silence while the UFO is observed. Two Ohio women, Angie Whitmeyer and Deborah Simmons, experienced this effect in 2001 while driving home one evening after a day of shopping in Dayton. The sky was crystal clear when they saw a bright light hovering over the nearby treetops. The two women parked the car on the side of the road to get a better look. As reported by Tim Swartz, "Deborah was shocked by how large and close the UFO was to them. 'The light was so bright and white that you couldn't see any shape behind it. But we could tell it was pretty big, at least as big as a house.' (...) Suddenly, another, identical bright light swooped down from the sky and hovered a short distance away. The two sisters decided the situation was becoming too strange and tried to drive away. 'That's when I discovered that the car had stopped and I couldn't restart it', Angie said. 'Nothing worked, the lights, the radio, it was completely dead.' The two women also noticed that an odd silence had descended over the area, accompanied by a strange feeling of isolation. Angie remembered that it seemed as if they were the only people in the world. 'I don't remember seeing another car come by during the entire time we were there, which is really weird because at that time of an evening there's always traffic on the road. It was just dead silent outside, no birds, nothing. It was as if we were in another world'."

"Uncertain what to do next, Angie and her sister continued to watch the strange pair of lights when, unexpectedly, both objects shot straight up and disappeared into the night sky. The area was plunged into darkness, and oddly enough, the normal sounds of the night came back almost as if switched on. 'As soon as the lights flew away', Deborah said, the car started running again all by itself. The lights and radio were on just as they were before everything happened'."

"According to their watches, the strange encounter had lasted more than 20 minutes. However, when they arrived home, Deborah's husband seemed unconcerned about what they thought was a late arrival. That's when they discovered that instead of being 9:oo p.m., as their wristwatches indicated, it was only 8:35 p.m. 'It was if the entire time we spent looking at those lights had never happened', Angie said. 'But it did happen, our watches showed that we had been stuck out here for over 20 minutes but somehow we gained that time back with a few minutes to spare. Normally, we should have been home at around ten to nine, but somehow, despite what had happened, we got there early'."(1)

In his article on the *time distortion* caused by UFOs, Swartz stated that experiences like those described above are not rare exceptions in the literature on UFOs. He mentioned the research by Mark Rodeghier of the Center for UFO Studies, who found that, when car engines stalled due to the effect of UFOs, "spontaneous engine restarts" occurred in about 10% of the cases as soon as the sighting of the UFO ended. The driver did not have to turn the key. According to Swartz, "one witness said: 'As soon as the UFO flew away, the car, radio, headlights, switched back on by themselves. One second everything was completely dead, and the next, everything was running smoothly as if nothing had ever happened.' The witness said he was left with the feeling that his car had been stopped between the ticks of a clock: 'Like time had completely vanished'."(2)

Swartz also described the experience of a deer hunter in Wisconsin who had observed a UFO hovering over the treetops directly over his head. "He stated that the day had been windy and the trees were swaying and creaking pretty loudly in the breeze. What made him look up was the fact that all of a sudden the forest 'went completely dead'. He noted that the trees had stopped moving as if 'frozen in place'. (...) As soon as the UFO passed overhead the forest returned to normal."(3)

The "frozen" sounds and motions associated with time distortions can be explained by the geometry of the encounters between abnormally large VE-spheres. As explained in the preceding chapter, the effects caused by such encounters differ, depending on whether they occur at the origin of an abnormally large VE-sphere or at its boundary. During an anomalous mind/matter interaction, the "normal" flow of time is changed by events involving abnormally large VE-spheres that are continuously re-created while the interaction occurs. Time is slowed at the *origin* of an abnormally large VE-sphere. This is the location where anomalous mind events are created by a strong focus of attention. It is where the abnormally large VE-sphere is continuously re-created. Time is speeded up at the *boundary* of the VE-sphere, which is where the expansion of the VE-sphere ends. At the boundary, the VE-sphere collapses in a cascade of encounters involving the QV-bubbles in its surface.

In the case of the two Ohio women who had pulled off the road to watch to UFOs, their wristwatches showed, as they said, that they "were stuck out there for over 20 minutes". Actually, much less time had elapsed. Their wristwatches were speeded up. Such "time warps" are consistent with Einstein's equations regarding the relativity of time and space. They show that if, hypothetically, a clock were suspended in outer space, far away from the gravitational field of any planet, it would tick faster. Time is speeded up in a weak gravitational field and is slowed in a strong gravitational field. UFOs cause anomalous distortions of time because they distort the gravitational field. I described the details of gravitational anomalies in chapter fifteen.

Let's turn to the process that speeds up the elapse of time at the landing sites of UFOs. In chapter sixteen, I explained that the process at a landing site is energized by the vacuum energy in the surface, or boundary, of abnormally large VE-spheres. The process generates an increase in temperature and scorches the earth. Later examinations of the landing sites showed that the plants at those locations had *aged more rapidly* than normal.

The process of premature aging corresponds to an accelerated decay of particles. Through a process that is explained in more detail in Part Three of this book, the configurations of interacting VE-spheres that are the particles called hadrons are kept in existence through acts of observation, which are encounters between VE-spheres. The encounters occur in morphic resonance and re-create the particles. The probability that such encounters will continue to occur depends on the number and the size of VE-spheres interacting at a specific location. If the number of encounters decreases, the rate of particle decay increases and time is speeded up.

This is what happens at the boundary of an abnormally large VE-sphere. If the top layer of the earth at the landing site of a UFO becomes scorched, this also affects the layer of earth underneath, which lies just *outside* the boundary of the anomalous physical system. Because of the cascade of encounters that occurs at the boundary of a VE-sphere, the presence of the UFO reduces the size of the QV-bubbles in the space that contains the scorched earth. But the volume of that space, obviously, remains unchanged.

If we visualize the bubbles as balloons that have become smaller, we can see that now there is more room for some of the expanding surfaces of the balloons in the adjoining space to expanding into the partially "vacated" volume of space. In other words: if the bubbles in one of two adjoining volumes of space become smaller, the bubbles in the other volume of space have more room to expand and become larger. This volume of space contains the roots of the plants that age faster. The decay of the particles in the physical substance of the roots is accelerated.

A striking example of an accelerated aging process is what reportedly happened to a man who was apparently influenced by the boundary effects of a UFO for an unusually long time. He had taken a walk in the woods in the evening, he later said, and had observed a UFO. When he did not return from his walk, his friends began to worry. He remained missing for several hours. When he finally did show up, confused and disoriented, he had a three-day growth of beard. He could not recall where he had spent the time.

If the story is true, the explanation is like the one proposed above. The man's mind was influenced by the fluctuating vacuum energy in what I have called *VE-space*. The biological events in his body were influenced by the corresponding quantum mechanical events in what I have called *QM-space*. They were influenced by the boundary effects of the UFO. The accelerated aging process in the man's body showed itself in the growth of his beard.

The opposite effect of UFOs, a slowing of time, involves what has been called "missing time". That event occurs when someone has a vivid experience involving a so-called close encounter with a UFO, which is usually followed by an encounter with "aliens". As described by those who had such an experience, the aliens paralyze their victims and abduct them into their spaceship, subjecting them to a brief but thorough examination and other procedures. The intense and deeply unsettling experience seems to last only a few minutes. But when the experience ends, the person who went through the ordeal discovers that much more time has elapsed. Sometimes, several hours "are missing".

During this type of experience, the person is in an altered state of consciousness. This is like a trance triggered by the focus of the mind on a baffling phenomenon. The mind is sometimes totally unaware of outside reality. Amplified by the energy of the collective unconscious, the mind contributes to the creation of the UFO it observes. The physical process that occurs during this anomalous mind/matter event is, in principle, the same one that occurs when persons who act as a "medium" in a séance create observable physical events. When they want a table to rise from the floor, they focus their mind on the table and "connect with it". Their mind, powerfully supported by the collective unconscious, counteracts the normal influence of gravity.

In chapter thirteen, I explained why a young girl named Kathleen briefly *gained weight* as the table began to rise. I also mentioned that she *lost weight* while energizing an undetectable source that produced rapping sounds on the table. The underlying process involves the phenomenon of time. The sound waves coming from the table take time to reach the person who is hearing them. Some of the energy is lost in transmission. But the undetectable influence that causes the rapping sounds reaches the table instantly. It is the instantaneous expansion of VE-spheres created by the mind events of the medium. This difference in time is the reason why there is a continuous outflow of vacuum energy from the medium to the table as long as the rapping sounds continue.

The weight loss is the gravity equivalent of the VE-surfaces originating in the mind of the medium. The VE-surfaces are bio-photons transmitting gravity effects from the medium to the table. Bio-photons connect the two locations directly. No other atoms (of a measuring instrument, for example) absorb these bio-photons. The bio-photons are part of a wave that transmits radiated energy. The rapping sounds, on the other hand, are part of a wave that transmits molecular energy. The molecules of the air vibrate as they transmit the rapping sounds through space, from the table to the eardrums of the person who hears the sounds.

The sounds that come from the table are produced by the undetectable vibrations of which Pribram and Bohm said (16/13) that they create the "hardened reality" of physical events. As I explained in chapter sixteen, that reality can be the mass (the configuration of gravity effects) that produces sounds by impacting on a table. Holographic images created by the mind are not necessarily the ghost-like figures seen by participants in a séance. They can be the image that exists in one person's mind and is not detectable by any other person.

Research into the processes associated with dreams supports the scenario of time discrepancies proposed here. Experiments have shown that many dreams last only a few seconds or a few minutes. But when participants in the experiments were asked to describe their dream, many of them said that it seemed to last several hours. The wide variety of events experienced in the dream could not possibly have occurred in a few seconds of measurable time.

During a dream, time flows much faster than normal. In terms of the number and apparent duration of events, a dream "covers a lot of ground".

Researchers can establish how long a dream lasts because certain measurable events occur as soon as a dream begins. Among them are rapid movements of the eyeballs while the eyes are closed, and electrical activities in the brain. These events no longer occur when the dream has ended. Participants in such experiments are awakened as soon as the instruments indicate that the dream is over, and are asked to describe it.

Dreams are energized by realities that exist *inside* the mind. They are pent-up emotions, thoughts, and memories. The physical effects of UFOs are energized by a reality that exists *outside* the mind. The observation of a UFO is an intense experience, "more real than ordinary reality". A dream, on the other hand, is usually less intense than a normal experience. As explained at the beginning of this chapter, abnormally intense experiences involve *fewer* differentiated mind events within a given period of time. Time is slowed. Less intense experiences can "cover more ground" and involve *more* events. Time is speeded up.

In contrast to normal dreams, prophetic dreams, which are like a glance into the future, or similar dreams energized by the collective unconscious, are intense experiences. The same goes for all experiences energized by abnormally large amounts of the vacuum energy in VE-space, which contains the collective unconscious. A Near-Death-Experience (NDE) is the most striking example of the process that creates an intense experience. It "compresses" time. I described the process in chapter fourteen. Depending on whether the anomalous events occur at the *origin* of an abnormally large VE-sphere, or at its *boundary*, time is either slowed or speeded up. The mind floating above the clinically dead body is at the origin of the VE-sphere. The clinically dead body observed by the out-of-body mind is located at the boundary of the VE-sphere.

Speeding up time is like storing a lot of information on a very small chip of the kind used in MP-3 players such as iPods. Such a chip can store hundreds of hours of music. On the chip, the music is stored like events that are the "substance" of time. Similar chips in digital cameras can store thousands of images. To look at all of them, one after another, the way we watch "moving pictures" projected from a reel of film, would take hours. During an NDE, the information contained in light waves is stored in the atoms of a dysfunctional brain while it still contains the potential energy that is needed if it is to be brought back to life.

As in all light waves, atoms interacting by exchanging photons transform the offer waves of light into echo waves, and vice versa. The photons that the atoms of the clinically dead body send out as offer waves are absorbed by the atoms at the location of the out-of-body mind. The corresponding echo waves from the location of the out-of-body mind carry the information back to the atoms of the brain in the clinically dead body. There, the information is stored like the compressed information in a digital chip. It is projected as a holographic image when the brain starts functioning again. The process is the same as the one that occurs in a living brain, as described by the neurophysiologist Karl Pribram. The wave patterns that are stored in the brain, he said, "can be retrieved later through the appropriate patterns of wave frequencies".(4)

What happens when the brain is brought back to life after a Near-Death-Experience is like the process that occurs during a dream and compresses the flow of time in the mind. When a dream occurs, the actual process that creates the dream may take only a few seconds, but the events that occur in the dream are experienced like a much longer period of time. In the scenario proposed here, the information stored in the brain during an NDE can be retrieved in a split second when the mind begins functioning again. The information includes all events that the mind observed from an out-of-body location during the time it took to bring the clinically dead body back to life. When the out-of-body mind has "returned" into the body and begins functioning again at the location of the brain, it experiences events that have already happened but were stored in the brain like memories.

When these "memories" are experienced (paradoxically, for the first time), the corresponding brain events occur almost instantly. In that respect, NDEs are like the experience during the observation of a UFO. In one of the incidents described by Swartz (17/2), the observer of a UFO said that it seemed as if his entire UFO experience had occurred "between the ticks of a clock". How fast does a clock tick? The clock used by physicists for a very precise measure of time is an "atomic clock". It makes use of the fact that the atoms in certain molecules vibrate at a constant frequency. The vibration is transmitted to an oscillator, which drives a clock. The vibration of atoms is pretty fast "ticking". The intervals are extremely brief. Nevertheless, based on what is known about vacuum fluctuations, the Physics of encounter suggests that a wealth of mind events can be crowded into such brief intervals. The vacuum fluctuations of mind events occur in the nuclei of interacting atoms.

During the observation of a UFO, the mind of the observer no longer contains the normal mix of "inside" and "outside" reality. The elapse of time that is the normal reality outside the observer's mind may no longer be experienced at all. "Outside time" may come to a complete stop. This happened to the deer hunter whose experience was described by Swartz (17/3). When the hunter saw a UFO in a forest during a strong wind, the branches of trees that were "swaying and creaking loudly" suddenly stopped moving and the forest "went completely dead". But the elapse of time inside the deer hunter's mind did not stop. He continued observing the UFO and was aware of his own existence as an observer. The hunter was seeing images created by what I have called the gluonic light inside the atoms of his brain, but his mind had shut out the input of what I have called photonic light. If time had also stopped inside his mind, he would no longer have been aware of anything. That would have been the same as falling into a coma, or losing consciousness.

Many UFO events involve experiences like those of the deer hunter in Wisconsin. The observation of a UFO is like a dream. The interpretation of dreams proposed here shows that time is not a uniform reality. There is an "unhardened", non-measurable aspect of time that exists inside the mind. It involves the vibrational energy of quantum vacuum fluctuations. The "hardened", measurable aspect of time exists outside the mind. The non-measurable reality of time inside the mind continues to "flow" as long as a person is conscious (or dreaming), even when the measurable reality of time outside the mind is no longer experienced.

To observers of a UFO, all motion outside their mind may appear to be frozen in time, but this anomalous experience occurs because their mind stops observing the motion and focuses on observing the UFO. What they see of the reality outside their mind is like a continued look at a snapshot of what was happening at the instant their mind turned its attention inward and shut out all further observation of the events occurring outside their mind. The motionless image of the reality outside their mind is like the background that fills a TV screen while the newscaster is delivering the important information "upfront". Normal awareness is a variable mix of events that originate outside and inside the mind. Sometimes the events that are "upfront" originate inside the mind, and sometimes the source of the experienced events is located outside the mind.

The realities inside and outside the mind are two physical realities that interact, or become "entangled". I have already mentioned that a limited degree of entanglement is a normal aspect of all physical processes that influence what we experience in our daily lives. The interaction creates extreme effects with respect to space and time, however, if it involves the events in a spaceship that accelerates to an "abnormal" speed, relative to what we normally experience on earth. Time is slowed inside the spaceship, relative to the time that elapses outside the spaceship. This means that passengers returning to earth after a high-speed excursion through space would have to deal with "missing time". The clocks on earth ticked faster than the clocks inside the spaceship. The passengers of the spaceship would discover that they aged more slowly than the people who remained on earth.

The effects of high speeds (more precisely: acceleration) can be calculated from the equations of Einstein's relativity theories and have been confirmed in experiments. But the effects that can actually be demonstrated are exceedingly small. Physical objects resist acceleration. Matter possesses inertia, which means it resists changes in motion. Acceleration increases the mass of matter. Energy is needed to increase mass through acceleration. The energy that would be needed to produce a large change in mass and time by far exceeds the possibilities of human engineering. The laws of physics that govern the creation of mass from energy prevent the transformation of such mathematical "thought experiments" into physical reality.

Before considering the effect of acceleration on the dimensions of space, and on the mass of an object, let's look at the effect on the dimension of time. I have already explained that, as Elitzur said (3/1), "there is more to time than just a dimension". *Time is a process.* The process creates not only the dimension of time. It also creates the dimensions of space. The process creates physical reality.

The process, briefly summarized, consists of interacting VE-spheres. Mass is generated by the Higgs field that is created where the surfaces of VE-spheres expand toward a point of encounter. Time is generated by the interaction between the QV-bubbles in these surfaces. That process creates the elementary ingredients of consciousness. They are the *qualia* that are created by oscillating strings inside the quantum vacuum of QV-bubbles. Qualia are the colors, sounds, feelings, and other sensations that are experienced on the inside of a QV-bubble. QV-bubbles expand and collapse at points of encounter between the surfaces of VE-spheres. The interaction creates new QV-bubbles that expand from the points of encounter.

It is logically impossible for consciousness to consist of an unending sequence of qualia that are all the same. If that were the case, you would not be conscious. It would be like looking at an eternally blue and empty sky. You would not be aware of your own existence as something different from outside reality. You experience qualia because they are *differentiated* sensations. Time elapses inside your mind and you experience your own existence when different qualia are continuously created and you experience all of them as your own sensations.

Experienced time is slowed inside an accelerating spaceship because the motion of the spaceship increases the number of encounters between the QV-bubbles of a passenger's consciousness with the QV-bubbles in the quantum vacuum fluctuations of space. This increases the probability that the motion will create a larger number identical QV-bubbles. The creation of identical QV-bubbles is a matter of chance. Encounters occur only in morphic resonance. The QV-bubbles that are created by encounters along the *same line of motion* are identical. For reasons explained in chapter three, differentiated QV-bubbles are created when successive encounters involve lines of thrust that occur *at right angles* to one another. They are the thrusts of oscillating strings.

The motion of an accelerating spaceship, at any one point in the line of travel, occurs along one dimension of space. If the spaceship were traveling at the speed of light, all qualia in the mind of a person traveling in that spaceship would be identical. At the speed of light, "time stands still". In the scenario of interacting QV-bubbles and VE-spheres, the probability of creating identical qualia corresponds to mathematical *probabilities* described by the theory of *quantum mechanics*. The expanding surfaces of VE-spheres are probability waves. In the proposed scenario, the interaction of VE-spheres also explains the effects described by the equations of *relativity theories*, which show that time and space can be compressed or stretched (dilated), depending on the motion of the observer. The Physics of Encounter links pivotal concepts of theories that seemed incompatible.

Einstein's equations quantifying the relativity of space and time show that an accelerated spaceship "shrinks". To visualize that process, think of the length of the spaceship as a straight line. At the speed of light, that length would be reduced to an infinitely dense point. The width and height of the spaceship would not be reduced, because those two dimensions are at right angles to the line of travel. But since the atoms of the spaceship located in those two dimensions would travel forward with the spaceship, they would be reduced to points located in those two dimensions of space. The spaceship would be compressed into something like a flattened tin can of zero-dimensional thickness. It would be like an invisible surface.

That surface would be the equivalent of a tiny segment in the surface of a VE-sphere that encompasses the entire energy of the universe. It is only a hypothetical surface because spaceships cannot attain a speed that is anywhere near the speed of light. But the image of such surfaces is in keeping with Einstein's *Gedankenexperiment* (imagined experiment) mentioned in chapter one. The surfaces of interacting VE-spheres contain the essence of time, space, and consciousness.

*NOTES*
(1) Swartz (2007), p. 24
(2) Ibid., p. 25
(3) Ibid., p. 25
(4) Stahlhut (1993), p. 16

# Chapter 18

The Universal Mind. Encounters across the barrier of time. "Ancient astronauts" and the spaceships of Indra and Arjuna. Archaeological artifacts and the future of humanity.

*Each of us is the Universal Mind but inflicted with limitations that obscure all but a tiny fraction of its aspects and properties.*
Henry Margenau

The intriguing aspects of time and consciousness highlighted in the preceding chapter are most strikingly apparent in a Near-Death-Experience. During an NDE, the human mind hovers at the boundary between observable reality and the hidden reality that contains what the psychiatrist Carl Gustav Jung called mankind's *collective unconscious*. The Physics of Encounter suggests that this hidden reality exists in the quantum vacuum fluctuations of space. The mind events stored in these timeless, undetectable fluctuations of potential energy are more than just passive memories. Together, they are an indestructible reality that various philosophers and scientists have called a universal mind, a conscious universe, or a cosmic consciousness. In the rare instances of morphic resonance between the human mind and the fluctuations of vacuum energy in our universe, this resonance empowers the mind to overcome some of the limitations mentioned in the above aphorism.

In this chapter, I will use Margenau's concept of the "Universal Mind". The elementary mind/matter interactions described in this book are evidence that the human mind, when energized by the Universal Mind, can be a powerful influence on physical reality. The "flow of time", expressing itself as a flow of events, cuts across the two main currents in the history of philosophy and science: what is the primal energizing source of events, matter or mind? In a metaphorical sense, the flow of time can be accelerated by humans inspired by their visions of progress. Science provides inspirations because it has harnessed physical forces through technological ingenuity. But science has also recognized the sheer power of the mind, of ideas. In a very real sense, elementary matter events are influenced by mind events, and these can be amplified through morphic resonance with the Universal Mind.

What about the "speed of time" in a world of matter that brings forth ideas? Is time a product of a strictly materialistic (physical) process? This question was addressed by Robert Havemann, a professor of physical chemistry at the Humboldt University in the communist sector of Berlin during the cold war. Havemann's classes attracted large numbers of students but they irritated the ideological watchdogs of the communist regime. Havemann pointed out some important aspects of time that were not in accord with the dogmatic party doctrine of dialectical materialism.

Havemann was a political dissident, but his fame as a scientist protected him, for many years, from the iron grip of the communist bureaucrats. He argued that dialectical materialism, properly interpreted, was an excellent tool for scientific analysis. In one of his lectures, many of which were later printed by a publishing house in the free part of Berlin, he pointed out that "one hour today is not the same as an hour several billion years ago". This means, Havemann argued, that the "constant" speed of light is not constant throughout the history of the universe. If the "tempo of time" in the universe were suddenly to double, he said, we would not notice this. "To say that there is an absolute tempo of time makes no sense, it is a notion devoid of meaning. All the more meaningful, however, is the idea that the measure of time is linked to the processes that occur in nature. (...) Time for us is determined by the events with which time is filled, by the events we experience."(1)

To illustrate the point he was making, Havemann mentioned the "animals whose biological processes slow down in winter and who become alert and lively in the warm sun. We must assume that these animals periodically experience changes in how fast time elapses in the world that surrounds them. If the temperature drops, time speeds up for them. (...) Conversely, for a lizard lying on a hot rock in the intense heat of the sun, the mosquitoes flying through the air move very slowly. The world the lizard sees is like the proverbial land of milk and honey. When the lizard flicks its tongue into the air, the mosquitoes place themselves right on it."(2)

Havemann's lectures caused headaches for supporters of the reigning materialist philosophy in the East German dictatorship because they recognized that his reasoning was in line with the research of physicists published in the democracy of the "enemy countries". Dogmatic materialism just couldn't handle the seemingly contradictory aspects of quantum mechanics. The examples provided in the preceding chapters show that the dialectical approach to the complex issues of time, mind, and matter can unify significant aspects of materialistic and idealistic philosophies. Time elapses only in what Karl Pribram (16/13) described as the reality that is "hardened" by the action of the mind. "Time does not enter the picture", as

John Briggs put it (1/2), "until a quantum system interacts with its surroundings". Time is the reality in which acts of observation contribute to the collapse of quantum mechanical wave functions and to the creation of elementary particles.

Quantum mechanical wave functions are the expanding surfaces of VE-spheres. Interacting VE-spheres, expanding and collapsing at points of encounter, are the quantum vacuum fluctuations of space. These unobservable events occur in the *timeless reality* of what I have called *VE-space*. The vacuum energy in VE-space is released at points of encounter and creates quantum mechanical events. They occur in what I have called *QM-space* and are an observable (measurable) reality.

There is an "unhardened", non-measurable aspect of time that exists *inside* the mind. It involves the boundless energy of the quantum vacuum that exists throughout space, including the vacuum fluctuations in an atomic nucleus. That is where the undetectable gluonic light of consciousness is created. The "hardened", measurable aspect of time exists *outside* the mind. Time continues to elapse inside the mind as long as a person is conscious (or dreaming), irrespective of whether the mind is aware of events outside the mind. The elapse of time *inside* the mind is generated by a timeless process *outside* the mind that creates the physical events in the atoms of the brain. That is a pretty good example of a dialectical process during which mind and matter interact on an equal footing.

During normal mind events, the expanding and collapsing VE-spheres that create the experience of time are of normal size. The quantum mechanical wave functions that are the expanding surfaces of VE-spheres "harden" vibrational energy into physical objects while the interaction of oscillating strings create the qualia of consciousness. During anomalous mind events that involve distortions of experienced time, the VE-spheres are abnormally large. But what is the "normal" size of a VE-sphere? Since the interaction of VE-spheres generates time, the answer to that question was given by Havemann (18/1): There is no way we can know the "absolute tempo of time (...). Time for us is determined by the events with which time is filled, by the events we experience."

The mind influences not only what Havemann called the tempo of time, but also the direction of what has been called the "*arrow of time*". As we experience the events in our universe, time only flows forward, from a cause to an effect that occurs later in time. But in the world of quantum physics, where particles appear and disappear unpredictably, *forward causation* and *backward causation* are equally possible. In Wheeler's view, for example, as described by Folger (7/9 and 7/10), "most of our universe consists of huge clouds of uncertainty" and contains "realms where the past has not yet been fixed". Our universe "is built like an enormous feedback loop (...) in which we contribute to the ongoing creation of not just the present and the future but to the past as well."

As Atmanspacher explained (10/12), the "symmetry of causation with respect to both directions of time" is broken by mind events that are created by quantum mechanical interactions involving the atoms of the brain. Such mind events are interconnected in a continuity of stable configurations, thereby assuring the existence of a personal identity and what Atmanspacher called stable "representations of reality" in the mind. Stable mind events are created in "hardened" physical reality as described by Pribram, and not in "clouds of uncertainty". In our normal, everyday experience of physical objects, therefore, time only flows forward. For the human mind, forward causation is, as Atmanspacher put it, "a selection criterion for stability".

During anomalous mind/matter interactions, however, as in precognition, someone may know in advance what will happen in the future, or have a prophetic dream. In those cases, the mind receives information about events that have not yet occurred. The influence of future events, it seems, flows backward in time and carries information directly into the mind. But is that the only possible conclusion? Could it be that the "extended mind", as described in chapter five, extends not only into space but also into the dimension of time? If our mind extends into the future and exists there, as well as in the present, it could instantly see a future event. It would not be necessary for an influence to flow backward in time.

The Physics of Encounter suggests that the effects *experienced* by the mind occur at points of encounter with the instantly expanding surfaces of VE-spheres. The effects *created* by the mind at points of encounter are also transmitted through space and time by the instantly expanding surfaces of VE-spheres. Space and time are created and continuously re-created by the interactions between VE-spheres expanding and collapsing at points of encounter in what I have called VE-space. The information carried directly into the mind is contained in the quantum vacuum fluctuations of the QV-bubbles expanding and collapsing in the surfaces of interacting VE-spheres. In VE-space, the mind instantly *receives* information from the future, or the past, through encounters that occur in morphic resonance. It also *transmits* information instantly through the same process. The big hurdle for these anomalous events is the fact that encounters between VE-spheres occur only in morphic resonance.

Through the interaction between VE-spheres, information can be instantly received and transmitted across space and time because, as Briggs explained, "time does not enter the picture until a quantum system interacts with its surroundings". Events that are fluctuations of vacuum energy in *VE-space* are timeless events. Time "enters the picture" in *QM-space*, in which quantum mechanical events are created at interrelated points of encounter. They are part of the "stable neuronal activities" in the brain described by Atmanspacher.

In the Physics of Encounter, the brain activities that determine the *location of the observer* at a specific instant are events in the nucleus of an atom. That instant is a point in space and time where an encounter between the surfaces of interacting VE-spheres creates the mind event of the observer. The interaction of QV-bubbles in the surfaces of VE-spheres creates the qualia of consciousness. A mind event straddles atoms that interact through the exchange of photons. It is a timeless event in a light wave. A mind event, therefore, is not just one point. It consists of countless points in space and time. The focus of consciousness, however, is within the "stable neuronal activities" of the brain.

David Lorimer, in a book about Near-Death-Experiences and "interconnectedness", underscored a viewpoint by F.C.S. Schiller that supports the above scenario. "Matter is not what *produces* consciousness but what *limits* it and confines its intensity within certain limits: material consciousness does not construct consciousness out of arrangements of atoms, but contracts its manifestations within the sphere which it permits." (Italics in original text.) This corresponds to the argument made by Eccles (13/1) that the mind exists as an independent reality and is not identical with the brain. Similar arguments were made by Henri Bergson and William James who, according to Lorimer, also "take issue with the conventional view that consciousness is produced by the brain. Bergson defines the brain as the organ of attention to life, an organ which channels, limits and focuses our awareness on the practicalities of the present moment." Bergson, Schiller, and James all suggest, Lorimer pointed out, that only "consciousness of a certain kind is permitted or transmitted through the brain".(3 )

"Consciousness of a certain kind", as interpreted by the Physics of Encounter, is a focused excerpt of the collective unconscious. The focus is provided by the brain for the individual human mind. The collective unconscious exists in the quantum vacuum fluctuations of space upon which, as Puthoff argued (7/2), "coherent patterns can be written". The coherent patterns are the memories of all of humanity that have accumulated through the ages. They include the mind events of all humans who are alive today and of all humans who are yet to be born. The coherent patterns, therefore, include every imaginable combination of the elementary ingredients of consciousness (qualia). The reservoir of energy in the universe is a reservoir of consciousness. The memory of mankind, or collective unconscious, is encoded in the timeless fluctuations of vacuum energy that occur in what I have called VE-space.

Bergson argued, according to Lorimer, that "the brain provides a framework" into which a specific memory can fit, but that it does not provide the memory itself. Bergson "concludes that it is a fallacy to make the automatic assumption that memories are actually stored in the brain".(4) In the scenario proposed by the Physics of Encounter, many of our memories *are* stored in the brain. They are memories created by our everyday experiences. But through morphic resonance, memories stored in the collective unconscious may influence our mind. These "memories" are events experienced by others. They can be a past event or a future event, or a distant event occurring elsewhere at the very moment when we, too, are experiencing it. The Physics of Encounter suggests that *all events*, not just mind events, are encoded in the collective unconscious as timeless patterns of fluctuating vacuum energy, whether or not they are experienced by individual human beings. They are part of the elementary constituents of consciousness in the Universal Mind that influences physical reality.

Lorimer's book on Near-Death-Experiences touches upon what could be called the collapse of time into a single moment of awareness. This happened, for example, to a mountain climber who had lost his footing and tumbled down a very high and rocky cliff. He felt sure that the end of his life was just a second or two away, but was miraculously saved from death because his fall was cushioned by some shrubbery. He later said that while he was falling, his mind saw every detail of his entire life, like images projected from a reel of film that was rolling at an incredible speed. Researchers into anomalous mind events use to term *"panoramic memory"* to describe this type of experience, because it is like looking at a panorama in space that can be seen at a glance.

According to Lorimer, panoramic memory was regarded by Bergson "as an indication that the past is indeed conserved. Extending his point of view", Lorimer wrote, "one might conjecture that moments of life-threatening danger focus our attention away from continuing material life as we begin to find our bearings in another dimension of consciousness, a wider and more intense state which some researchers have tried to explain by using the model of 'hyperspace'. Gordon Greene, for example, postulates that time is spatialized into a fourth dimension which enables us to perceive the whole of our past life simultaneously."**(5)**

Greene's argument supports the scenario proposed by the Physics of Encounter. The concept of "hyperspace" corresponds to what I have called VE-space. *Time is "spatialized"* in the sense of being a part of VE-space, which presents itself instantly to a hypothetical observer who is "pure mind" and no longer burdened by a physical body. VE-space consists of the geometric configurations created by interacting VE-spheres. It is a space of curved (spherical) surfaces encountering one another and instantly creating other curved surfaces. As Wheeler said (2/1), "there is nothing in the world except empty, curved space. Physics is geometry." The encounters between instantly expanding VE-spheres create time by creating differentiated ingredients of consciousness (qualia) in VE-space while creating quantum mechanical events in QM-space. Elementary particles are configurations of encounters and are kept in existence by acts of observation. A particle decays if acts of observation no longer occur in the required configuration. The decay of particles is an objective measure of the elapse of time.

The dialectical scenario that the timeless events in VE-space create the time that elapses in QM-space is supported by Wheeler's argument, mentioned above, that the universe is like an enormous feedback loop "in which we contribute to the ongoing creation of not just the present and the future but to the past as well." The present time is the time that "presents itself" to our consciousness while we are alive. A mind that is detached from a physical body, like the mind that floats above a clinically dead body during a Near-Death-Experience, can give us an inkling of what time "really" is. In the timeless process that creates time, everything *is*.

We can assume, therefore (but cannot prove), that our universe has no beginning in time and will not cease to exist. For the embattled defenders of dialectical materialism, this has always been the preferred conclusion. The Physics of Encounter suggests that the "big bang" was not and will not be the only explosive event in the history the universe, and that the universe will not expand into nothingness. The universe expands and collapses like the VE-spheres that are its elementary, unobservable "substance". This scenario resonates with the philosophy of dialectical materialism, but its supporters will have to face the facts brought to light by modern science, which indicate that the mind is a proud reality of its own and not subservient to the role of matter. The minds of humans inhabiting our planet are not like islands of consciousness that have emerged from a sea of lifeless atoms. The mind contributes to the creation of what is. The eternal reality of the mind is not a fluke of evolution and of quantum mechanical probabilities.

In chapter eleven I mentioned that the theologian and poet Ernesto Cardenal, in his collected works entitled *Cantico Cosmico* (the Cosmic Song), spoke of a pulsating universe, of expansion and never-ending "collisions" creating new events. He said that we are "atoms trying to understand atoms". We must understand the nature of what we are. We are joined together in a field of consciousness, aware of one another and of our own existence. Our mind, as part of living matter, straddles atoms interacting through the exchange of photons, which are surfaces of expanding possibilities. While photons are emitted and absorbed as quantum mechanical events in what I have called QM-space, the expanding surfaces of VE-spheres interconnect the nuclei of the interacting atoms. The interconnectedness is all-encompassing. The details revealed in the rapidly accumulating literature on *Near-Death-Experiences* speak volumes.

In the words of a Near-Death survivor mentioned in a book by Carol Zaleski, "light floods the mind, expanding awareness until it seems to comprehend everything in a single gaze".(6) Among the Near-Death-Experiences listed by Kenneth Ring are prophetic visions about the future of mankind, the sharing of thoughts with other disembodied minds, and a sense of rapture, joy, and magnificence.(7)

The strengthening of basic spiritual values through Near-Death-Experiences has far-reaching implications. The religious outlook is broadened and becomes more sophisticated. NDEs result in profound personality changes. Among the most important ones listed by Stahlhut are:

The fear of death disappears, combined with an unshakable conviction that consciousness survives physical death. Survivors speak of having experienced indescribable beauty and peace. Life is recognized as a cherished gift. Survivors feel imbued with empathy and compassion and a desire to dedicate their life to a purpose, to helping others. Sensitivity and the feeling of interconnectedness are heightened, material possessions become less important. Survivors feel that they have become a stern and incorruptible judge of their own past life, that there is no additional "punishment" waiting for them after death. During their Near-Death-Experience, they felt the sorrow and the pain they inflicted on others as their own deep sorrow and pain. They recognized their own imperfect humanity.(8)

The key concept for understanding the phenomenon of Near-Death-Experiences, and a wide variety of other anomalous experiences, is the concept of fluctuating vacuum energy, which can be harnessed, to a degree, during anomalous mind/matter interactions, as described in the preceding chapters. Through morphic resonance, the mind is put in touch with the timeless force of interconnectedness that fills the universe. This provides an explanation for the power of healing and of group consciousness, and also underlies the phenomenon that Carl Gustav Jung called *synchronicity*.

Synchronicity, as interpreted here, is an unpredictable and highly improbable coincidence seemingly based on "pure chance". It interconnects mind events and matter events across space and time, even across centuries. The link is established by morphic resonance. A striking example of Jungian synchronicity linking mind events with a physical reality in the distant future is provided by a religious text in the Hindu equivalent of our bible. It describes the battle between the God Krishna and the demon king Shalva, who descended upon the earth around 3200 B.C. UFO researchers supporting the extraterrestrial hypothesis have cited the text, written long after the legendary battle, as evidence that the earth was visited by *"ancient astronauts"*.

In the italicized part of the following sentences, I have pieced together my translations of the images used in the Sanskrit text according to a compilation published in German by Lutz Gentes.(9) The text is strikingly suggestive of 20th century warfare: *Fiery lances hurtled through the sky with thunderous noise, bursting meteors fell from the sky, and fiery balls propelled by expanding air split into many pieces with a thunderclap from billowing clouds.* After the attack, *calamitous whirlwinds swept through the city and the air was darkened with dust.*

The images evoke visions of the firestorms that swept through Rotterdam, Hamburg, and Dresden after the air raids in World War II, of rockets, bombs and artillery shells (propelled by "expanding air"). The Sanskrit text states that Krishna was attacked by a *city of steel moving freely through the sky and defended himself with high-flying arrows endowed with perspicacity that sought out the enemy by the sensing of sounds and shattered the enemy's weaponry into hundreds of pieces. The sky resounded with violent noise and seemed filled with hundreds of suns and countless stars.*

Could "science fiction writers" of pre-industrial times have conjured up such accurate visions of 20th century military technology? The use of an airplane, a flying "city of steel", or even the use of explosives in warfare was still centuries away.

The images of space travel in the above-mentioned religious text also seem to mirror today's realities: Indra, the God, *descended from the sky in a resplendent vehicle* and invited Arjuna, the son of king Pandu, to come aboard. Arjuna enclosed his body in *air-filled armor* and ascended in a craft that seemed like the *smokeless tip of a burning fire, filling all of space with a roar.* These descriptions bring to mind mankind's early ventures into space from Cape Canaveral. A space capsule was attached to the tip of a rocket ("a burning fire") and astronauts wore pressurized ("air-filled") spacesuits.

Also cited as evidence of extraterrestrial visitors in pre-historic times are archaeological artifacts. A stone tablet depicts a god who seems to be piloting a rocket. In other artifacts, we see figures who seem to be wearing spacesuits and helmets of the kind used by today's astronauts. The publications of Erich von Däniken contain a wealth of speculative interpretations along similar lines, based on what are undisputed archaeological anomalies. Remnants of structures testifying to engineering feats that were far ahead of their time and reappear in diverse cultural contexts and different epochs of human history challenge today's archaeological wisdom. Some of the structures seem to have been built either in expectation of visitors from outer space, or – as argued by some ufologists – by the extraterrestrial visitors themselves, who brought advanced technology to the civilizations that existed in former times.

Do these archaeological findings and the religious text quoted above support the assumption that the earth was visited by "ancient astronauts"? What source of information provided the creative minds of pre-industrial times with such accurate visions of future space travel and military technology?

It is not logical to attribute the archaeological findings and the events described in the religious text to God-like extraterrestrials. Why would God-like beings from other worlds use rocket propulsion, antiquated by today's science fiction standards, for thunderous maneuvers through space? The UFOs described by observers in our day and age are eerily silent spacecraft propelled by what are assumed to be anti-gravity devices. Why would aggressive visitors from outer space use bombs and missiles? The aliens encountered today paralyze their victims at will.

The above texts and archaeological findings are a strong indication that the phenomenon ascribed to "ancient astronauts" are Jungian synchronicities that transmit the contents of mind events by *retrocausation*, backward in time, to the creative minds of generations who lived before us. The synchronicities mirror what was literally burned into the minds of millions of soldiers and victims of war in our time. It stands to reason that the roaring rockets that signaled the awe-inspiring advent of space travel were a fitting image of an alien force in an age when thunder and lightning were regarded as the wrath of Gods. The war against Hitler's Germany, vastly more dramatic than any war that preceded it, as well as mankind's first breathtaking ventures into space, apparently made imprints on the minds of humans who lived long before us.

Synchronicities involving encounters across the barrier of time are random events that occur only in morphic resonance. If complex images are involved, the probability of encounters between the corresponding configurations of VE-spheres is low. A number of remarkable retrocausation effects did leave an imprint in our past. Among the evidence for these anomalous events are portions of many religious texts as well as the painted or sculptured images that were later discovered as archaeological artifacts. Since they are the products of extremely rare synchronicities, there might be no more. But the absence of traces in our past that point to a future beyond the primitive stage of today's space travel and warfare might be an ominous sign.

The alien "astronauts" of our times, according to those who profess to have encountered them during a UFO experience, have issued solemn warnings that we are endangering our survival by waging war and ravaging our environment. The traumatizing experience of what seemed to be an abduction by aliens have been described and analyzed by the Harvard psychiatrist John E. Mack. He interviewed people from all walks of life, examined them thoroughly and used only reports from people with an indisputably intact personality. Encounters with aliens, he concluded, are not symptoms of a mental disorder.**(10)**

The Physics of Encounter suggests that encounters with alien minds are fragmentary perceptions of a reality that is more elementary and more comprehensive than the one we observe in our everyday lives. The proposed model of mind/matter interaction provides arguments for the presence of mind events throughout space, allowing metaphysical speculations about transpersonal, disembodied minds, and about a life after the death of our physical body. Data supporting this conclusion have been compiled not only by the scientists who have investigated Near-Death-Experiences. Research into what has been interpreted as reincarnation also falls into this category. The process that deposited Napoleon's experiences into the mind of a woman in today's America, as described by Almeder (6/1), is a striking example of synchronicity. It cannot be considered as proof that a person is "reborn" in the body of someone else. It merely suggests that our mind survives the death of our body.

The process that transmitted Napoleon's memories across space and time into another person's mind is the same type of process that creates the insights and the encounters with disembodied minds during NDEs. It is this synchronicity, or morphic resonance, that triggers the encounters with "alien beings" during UFO experiences. All mind events are encoded in the timeless reality of quantum vacuum fluctuations, which can influence events in both directions of time. The minds of "aliens" are transpersonal minds that contain the wisdom acquired by future generations of humans. Their solemn warnings are voices from the future. Since we have to rely on morphic resonance for all of the anomalous experiences that provide us with glimpses of a reality that lies beyond our own life span, much of what we see during UFO experiences contains images of our past, in particular those that Jung called archetypes.

We are free in our efforts to determine what our future will be. It does not have to be as described by Chet Snow, who conducted experiments with hypnotic progression. Hypnotic states of consciousness can open our mind not only to memories buried in our own past, but also to information contained in the collective unconscious. Snow asked his hypnotized subjects to report what they experienced as their "future incarnations". Some of the subjects described how they experienced their own death, and said that they then found themselves alive again in a later period of human history. By the year 2300, according to these accounts, our planet will be depopulated, ravaged by drought, freezing rain, and widespread devastation. The experiences described included suffocation in a black cloud and death in an atomic explosion **(11)**.

We should not dismiss these apocalyptic visions by equating them with the current malaise about the course of events. In the various descriptions of mankind's future, the experience of one's own death was associated with the same paradoxical sensation of joy and empowerment that characterizes NDEs and the encounters with "aliens" communicating their concern.

We experience an enlarged wisdom within ourselves during the elusive moments when the mind events described in this book create a synchronicity, or resonance, with the creative force in our universe. The experience is a fusion with an all-encompassing consciousness. The wisdom that is passed on to us is contained in the experiences of those who have lived before us and in the future realities experienced by those who will be born after we have died.

The Universal Mind makes its presence known to us through our encounters with the minds of other persons. Egon Freiherr von Eickstedt has suggested that the word "person" may come from the Latin verb *per-sonare*: "Every person is like a sonic medium through which the eternal reality of our universe makes itself heard in the transient reality of our lives".**(12)** Since scientists are among the first to tell us that we should let the facts speak for themselves, the Physics of Encounter suggests that the wealth of anomalous phenomena that many of us experience in the transient reality of our lives contains a hidden mind/matter reality that can tell us something about the world in which we live. We should listen.

*NOTES*
(1) Havemann (1964), p. 69
(2) Ibid., p. 70
(3) Lorimer (1990), pp. 28-29
(4) Ibid., p. 29
(5) Ibid., p. 30
(6) Zaleski (1987), p. 124
(7) Ring (1984), pp. 142, 192, 219

(8)  Stahlhut (1994), pp. 1-15

(9)  Gentes (1996)

(10) Mack (1994)

(11) Snow (1991)

(12) von Eickstedt (1954), p. 116

# PART THREE
## Crucial issues in theoretical physics
### *Chapter 19*

**The hidden processes underlying current conceptual constructs.**

**Introduction.** The physical processes that create consciousness have not yet been adequately identified. The interaction between mind and matter is still poorly understood. The same holds true for many "ordinary" (not anomalous) physical phenomena. The Physics of Encounter suggests that in all of these cases, the underlying processes are the same.

I have written this part of the book because I want physicists, in particular, to recognize the merits of the proposed model of mind/matter interaction not only in terms of its relevance to anomalous phenomena, but also within the larger framework of fundamental theoretical issues. The following chapters provide strong evidence in support of the proposed scenario by showing that it dovetails with important aspects of mainstream theories and with broadly accepted experimental data. Included here, as additional supporting evidence, are a number of examples showing that the proposed scenario clarifies theoretical issues and contributes to a better understanding of current conceptual constructs.

For brevity, I will use some technical terms in this final part of the book. To readers who are not physicists let me say that the important thing is not to understand every detail but to recognize the diversity of intertwined data that must be understood in the context of anomalous phenomena. For the benefit of physicists, here is a synopsis of the Physics of Encounter, taken from my presentation to the *Society for Scientific Exploration* in 2005. I described the proposed scenario of mind/matter interaction as "the hypothesized relationship between the unobservable quantum vacuum fluctuations of space and the emergence of sentience from observable physical processes". This scenario "bridges the ontology gap between the physical reality of processes in the brain and the subjective reality of qualia". It "suggests that (...) quantum vacuum fluctuations can be geometrically described as a pre-manifest reality of expanding and collapsing spheres, interacting (...) at points of encounter between their surfaces. The spheres collapse at points of encounter and a new sphere expands from the point that absorbs the effects of the encounter. The effect simultaneously absorbed and exerted at that point (...) is either the origin of an elementary mind event or a participant in the creation of an elementary matter event, depending on whether the absorbed effect is experienced inside the sphere or exerted by the outside of its expanding surface. The proposed scenario shows that mind and matter can be assumed to arise, as emergent reality, from a common substratum of interacting, hidden effects."(1)

In this scenario, the point of encounter is the origin of what I have called a *QV-bubble*. It contains a *quantum vacuum* that expands and collapses as QV-bubbles interact. These interactions are the quantum vacuum fluctuations of space. For the above-mentioned spheres, I coined the term *VE-spheres* because their surfaces contain the *vacuum energy* in a proliferating number QV-bubbles that energize the expansion of the surfaces. As QV-bubbles expand, they become VE-spheres. For brevity, I have occasionally used the term VE-surfaces for the surfaces of VE-spheres. They correspond to the probability waves of quantum mechanics. The layers of proliferating QV-bubbles in the surfaces of VE-spheres are wave patterns, i.e. quantum mechanical superpositions. Encounters between these surfaces occur only if the waves are in resonance and cancel each other. This corresponds to what Sheldrake called *morphic resonance.*

The conceptual construct of *oscillating strings* plays an important role in this scenario. The surfaces of VE-spheres are undetectable entities that consist of interconnected points. The points are the endpoints of oscillating strings. The Physics of Encounter suggests that a QV-bubble is created when the thrusts of oscillating strings converge on a common point of encounter and cancel one another in all spatial dimensions, as explained in chapter three. The interaction of QV-bubbles creates the dimension of time. As interpreted here, time is a process. Time elapses as QV-bubbles expand and collapse at points of encounter. The vibrational patterns of interacting strings in the QV-bubbles of expanding VE-surfaces create the two manifestations of time: the elementary constituents of consciousness (continuously changing qualia) and the succession of quantum mechanical events that occurs because interacting QV-bubbles, i.e. the oscillating strings within them, create elementary particles.

The expansion of a VE-sphere toward a point of encounter is instantaneous. Since time is a process involving acts of observation, which are successive encounters between the expanding surfaces of VE-spheres, the expansion itself could be called "timeless". As part of the quantum vacuum fluctuations of space, the expansion is neither fast nor slow. Nothing actually moves. The coherence of the invisible surface of a VE-sphere is mediated by the interaction between the proliferating QV-bubbles in the surface. They contain identical qualia of consciousness. A measurement of time and speed is possible only with reference to what Atmanspacher called stable mental representations of reality in coupled map lattices.(2) These are mind/matter events that are created at interrelated points of encounter, as explained in chapter three.

There are many cases illustrating the correspondence between the proposed scenario and the concepts, theories and experimental data in the scientific literature. I have divided those that best exemplify the correspondence into three categories, of which each is presented in a separate chapter. This chapter highlights the simple, straightforward geometric aspects of the scenario: spherical surfaces, curved planes, and the straight-line thrusts of oscillating strings. The three subsections of this chapter contain information about the following three topics:

(a) the basic explanatory aspects of interacting spheres and hemispheres,

(b) the proposed interpretation of the EPR-paradox,

(c) the conceptual constructs of the string theory.

The three subsections in chapter twenty deal with:

(a) the process of photon emission and absorption,

(b) quantum gravity and dark matter,

(c) loop quantum gravity and angular momentum.

Chapter twenty-one proposes a classification of particles based on the Physics of Encounter.

Chapter twenty-two, finally, proposes some experiments to test the validity of the scenario described in this book. The experiments would show whether the hypothesized processes of mind/matter interaction influence the characteristics of particles as predicted.

**(a) - Spheres and hemispheres. The efficient cause and the final cause of quantum events. The wave/particle duality. The two halves of our universe and the "shadow brane".**

The interactions between VE-spheres correspond to what Lederman (10/6) called the "system configuration" through which particles acquire their mass. (See note in the next paragraph.) Chapter ten describes how the bosons in a Higgs field are created by interactions between the superposed layers of QV-bubbles in the surfaces of VE-spheres expanding toward a point of encounter. The process through which the interacting surfaces create particle spin and magnetic lines of force is described in chapter twelve. The configuration that creates the electric charge of particles will be described in chapter twenty-one.

>As in the previous chapters, the numbers in parenthesis and separated by a slash refer to the number of the chapter and the number of the reference note, respectively, where a quote was first mentioned and elaborated.<

Physicists have argued that in quantum mechanics, the collapse of a wave packet has an *efficient cause* and a *final cause*. The scenario described by Physics of Encounter clarifies the meaning of these complementary concepts. According to Richard Bierman, two observers are involved in the collapse of a wave packet, a "pre-observer and a final observer".(3) The wave packet is the expanding surface of a VE-sphere. It consists of the wave patterns created by the interacting QV-bubbles that expand and collapse in the surface of the VE-sphere. The effect of the observer is also a wave packet, i.e. an expanding VE-surface. It consists of the QV-bubbles that contain the qualia of the observer's consciousness.

When the two expanding VE-surfaces encounter each other in morphic resonance, both surfaces collapse and a new VE-sphere expands from the point of encounter. The hemispheres of that VE-sphere expand into opposite directions, each toward a point that may or may not become a new point of encounter. If either hemisphere is encountered, the other one also collapses. A VE-sphere collapses in its entirety if it is encountered at any one point. The expanding VE-surface is the *efficient* cause of a quantum mechanical event, because it is *certain* that the VE-surface will eventually be encountered and that the encounter will result in a quantum mechanical event. But the expanding surface is not the event itself. The *final* cause of the event is the encounter that creates a new VE-sphere. Since encounters occur only in morphic resonance, the final cause is a random event. Without morphic resonance, the expanding surfaces permeate each other and continue to expand.

In chapter five, I compared the hemisphere of an expanding VE-sphere to the smile of the Cheshire Cat encountered by Alice in Wonderland. The hemisphere is nothing by itself, but as it expands toward the observer in the dimension of time, it transmits the possibility of a future event, relative to the observer. Hemispheres that expand *toward* the observer transmit the effect of forward causation. Hemispheres that expand *away* from the observer transmit the effect of backward causation.

The proposed scenario applies not only to the world of quantum mechanics, but also to the universe as a whole. The astrophysicist Paul Davies suggested that the universe might consist of symmetrical halves. The boundary would be a temporal divide, he said, with time flowing forward on one side and flowing backward on the other side. Cycles of cosmic expansion and collapse would be separated by successive big bangs.**(4)** This corresponds to the cosmic scenario described by the Physics of Encounter.

As interpreted here, the big bang that created the universe was an encounter between two cosmic hemispheres, or "branes", in Stephen Hawking's metaphor, that expanded toward each other. The big bang created the cosmic sphere that is our present universe. It corresponds to what I have called a VE-sphere. The two hemispheres of this cosmic sphere are now expanding away from each other into opposite directions of time, relative to an observer of our universe. The hemispheres are the two "halves" of our universe described by Davies. The observer is located in one of the two hemispheres. It is the curved space of our universe. This is in keeping with Hawking's metaphorical image that "we live on a brane".**(5)** According to the science journalist Rüdiger Vaas (5/4), Hawking argued that the big bang was triggered by a pre-existing physical reality, and that this reality was shaped like a "four-dimensional hemisphere".

Hawking's use of the term "brane" corresponds to a hemisphere of an expanding VE-sphere. Because we "live on a brane", he argued, we cannot see the other hemisphere. The light we observe in the expanding hemisphere that is our universe does not travel to the other hemisphere. It corresponds to what Hawking called the "shadow brane".**(6)** In contrast to light, the gravity effects of the shadow brane, according to Hawking, do influence our universe. This, too, follows from the scenario proposed here. The details are described in chapter twenty.

Interactions between hemispheres that are the microcosmic equivalent of the "big bang" are quantum mechanical events and create what Michael Manthey (4/9) called "bit bangs", i.e. basic units of information for the conscious mind. The interaction between VE-spheres generates consciousness as well as gravitational effects. As described in chapter twenty, these effects accumulate as dark matter in the outer regions of the universe. They will eventually cause the gravitational collapse of our universe. A new big bang will follow.

In the unobservable physical reality of interacting VE-spheres, their expanding hemispheres create what James Beichler called the "reciprocal subspaces" of hyperdimensional theories.(7) As explained in section (c) of this chapter, string theorists have also argued that physical reality consists of additional, unobservable dimensions. The Physics of Encounter describes the process that creates these dimensions.

Michael Ibison, in his description of influences that are transmitted instantly, used the concept of "a moving now".(8) In the Physics of Encounter, the undetectable physical entity that "moves" through space is the instantly expanding surface of a VE-sphere. It contains the hidden energy for the creation of an elementary matter event as well as the qualia of consciousness that the observer experiences as "now". The "moving now" does not actually move. The expanding surface of a VE-sphere is continuously re-created at points of encounter between the proliferating QV-bubbles in the surface. The QV-bubbles are created where the thrusts of countless oscillating strings converge on a common point of encounter from all directions, as described in chapter three. An expanding VE-surface corresponds to what Ibison called a line of "zero-dimensional points moving through space".(9) The points are points of encounter between the QV-bubbles in an expanding VE-surface..

The Physics of Encounter clarifies the paradoxical wave/particle duality by showing that the interrelated points where acts of observation create quantized physical reality are located on interacting VE-surfaces. The expansion and collapse of these surfaces, and of the interacting QV-bubbles within them, are wave patterns. The oscillating surfaces delineate the boundary of particles that occupy space.

The wave/particle duality of photons is a special case because a photon, as a particle, does not occupy space. A photon, as a particle, is a point of encounter between the expanding surfaces of VE-spheres. A VE-sphere is the wave aspect of a photon. The surface of that VE-sphere impacts on the surface of an atom, which is a stable configuration of QV-bubbles in the surfaces of VE-spheres. The scenario of interacting VE-spheres clarifies the puzzling results of the double-slit experiment with photons. Even if only one QV-bubble in the expanding surface of a photon passes through one of the two slits, the VE-sphere is instantly re-created on the other side of the slit.

As in all cases of expanding VE-spheres, the energy for the re-creation is released at points of encounter with the hidden energy in the quantum vacuum fluctuations of space. The quantum vacuum fluctuations are expanding and collapsing QV-bubbles. A photon does not become observable energy until it interacts with a quantum mechanical reality. This occurs, for example, when the photon is absorbed into an atom. The surface characteristics of a photon traversing space as part of a wave are described in chapter twenty-one.

Most physicists agree that space is not a quantized physical reality, but that there is a need to define the smallest possible segment of space. It would have to encompass more than the one dimension attributed to the thrust of an oscillating string. To identify that unobservable "piece of empty space", physicists have introduced the concept of a *quaternion*. As interpreted here, a quaternion consists of four hemispheres, of which two expand toward a point of encounter and two expand away from the point of encounter. The convergence and the divergence occur simultaneously and along orthogonal axes. It corresponds to what string theorists have called a "flop transition". Section (c) of this chapter provides more details.

**(b) - The EPR-paradox and non-local correlations. Bohm's "hidden variable". Opposite spin effects at two points on a VE-surface. Collapsing superpositions.**

The examples provided in this chapter show that a large spectrum of "normal" but puzzling phenomena can be explained by a scenario that combines the conceptual constructs of oscillating strings and of interacting VE-spheres. This is strikingly apparent when the EPR-paradox is examined in the light of the proposed scenario. The paradox straddles the domain of mainstream physics and the research into paranormal phenomena.

The EPR-paradox was predicted by Einstein, Podolsky, and Rosen in 1935 **(10)** and has since been experimentally verified. It appears to involve an instantaneous communication between two particles manifesting a correlated behavior. One particle instantly "knows" about the creation of a twin particle. The two particles simultaneously observed at the instant of their creation can be thousands of miles apart, but if one has a left-handed spin, the other one exhibits a right-handed spin, and vice versa. This appears to violate the principle that no signal can traverse space faster than the speed of light.

The creation of the two particles with correlated characteristics was attributed to a "hidden variable" by Bohm *et al* **(11)**, but the process cannot be satisfactorily described by the currently known laws of physics. In the scenario proposed here, the hidden variable is the surface of an instantly expanding VE-sphere. The creation of the twin particles at the same instant and at widely separated locations appears to be a correlation between two *separate* events, but physicists assume that the correlated events are actually two manifestations of a larger, *single* event. The Physics of Encounter proposes the following description of this event.

The instantly expanding surface of a VE-sphere that is the hidden variable mentioned above is energized by the effects of oscillating strings. They create layers of QV-bubbles in the expanding surface. I described the process in chapter three. The layers of QV-bubbles are superpositions in the surface of the VE-sphere. The surface is a probability wave. The expansion and collapse of the interacting QV-bubbles in the surface are wave patterns. The expanding and collapsing quantum vacuum in the interacting QV-bubbles are the quantum vacuum fluctuations of space.

Where the surfaces of VE-spheres permeate each other, the oscillation of strings creates the spin effects of particles. I described the process in chapter twelve. In the proposed scenario, an act of observation occurs at a point of encounter between the surfaces of VE-spheres, if one of the surfaces contains the mind event of the observer. That mind event occurs when the encounter creates the qualia of the observer's consciousness in the QV-bubble that expands from the point of encounter and becomes a new VE-sphere. In the surface of that VE-sphere, every QV-bubble in the same layer of QV-bubbles contains identical qualia.

In the theory of quantum mechanics, the physical system that exists before the observation of an EPR-effect is assumed to be a superposition of conditions that collapse in their entirety when they are observed, or measured, at any one point.**(12)** In the scenario of interacting VE-spheres, the superposed conditions that collapse exist in the outermost layer of QV-bubbles that energize the expanding surface of a VE-sphere. When that surface is encountered at any one point, the encounter is an act of observation, or a measurement. The event involves two QV-bubbles. One of them is the QV-bubble in the surface of the VE-sphere and the other one is the QV-bubble that contains the qualia of the observer's consciousness. The event creates a new QV-bubble that expands into a new VE-sphere.

As explained in chapter three, the creation of QV-bubbles is energized by oscillating strings, and these vibrational patterns are the qualia of consciousness. The vibrational patterns correspond to the mind event that occurs during the observation of an elementary particle. The physicist Greene, in his description of the process (3/7), explained that "what appear to be (...) particles are actually different 'notes' on a fundamental string".

The EPR-paradox occurs when two quantum mechanical events are created by simultaneous acts of observation (measurements) at two separate locations, as agreed upon by the physicists who want to confirm the EPR-paradox in their experiment. Each measurement is an encounter with the expanding surface of a VE-sphere, and each measurement creates a new VE-sphere that expands from the point of encounter. Each measurement is influenced by the effect of the instantly expanding surface of the VE-sphere that is created by the other measurement.

If we designate the two points of encounter as points A and B, we can say that one hemisphere of the VE-sphere that is created at point A expands toward point B, and vice versa. The two expanding hemispheres permeate each other and create spin effects, as described in chapter twelve. The spin effects observed at the two points of encounter are opposite because they are created by the oscillating effects in surfaces that originate on opposite sides of the same VE-sphere. They are like mirror images of each other.

The vibrational pattern of the probability wave that contributes to the creation of a quantum mechanical event at each location of the two simultaneous measurements is a superposition, or layer of QV-bubbles, in the expanding surface of a VE-sphere. Since all QV-bubbles in a superposition are identical, the QV-bubble that participates in either of the encounters at the two locations contributes to the creation of twin particles with opposite spin.

There is no limit to the size of VE-spheres interacting in morphic resonance at points of encounter. The VE-spheres can attain intergalactic size and can be as small as the wave patterns of high-energy photons, or even smaller. VE-spheres are photons if their surfaces are energized by four effect planes, as described in chapter twenty-one. This is only the case if the entirety of their effects is absorbed into atoms interacting within a light wave. The VE-spheres involved in the EPR-paradox, however, are not absorbed in their entirety. Only superpositions within their surfaces interact with the observers at separate locations and are absorbed into atoms. The superpositions contain the potential energy of photons. Photons are smaller VE-spheres located in the surfaces of larger VE-spheres.

The EPR-effect is a *"non-local"* correlation because the process that creates the correlation is an undetectable physical process. It has no definable location in space. The two correlated events through which the EPR-effect manifests itself may occur anywhere. The interacting surfaces of VE-spheres, or probability waves, are the quantum vacuum fluctuations of space. They are what Puthoff called a pre-manifest reality. The instantaneous correlation of widely separated events results from the morphic resonance between the instantly expanding surfaces of VE-spheres interacting at points of encounter.

**(c) - String theory. "Flop transitions". Sub-Planck distances and the Kaluza-Klein radius. Undetectable dimensions. Loops and one-brane configurations.**

Brian Greene described one of the configurations that result from the interaction of oscillating strings as a "space-tearing flop transition". It occurs where space "pinches down to a point" and then "expands in a new way".(13) This corresponds to the interaction between expanding VE-spheres. The spheres expand from QV-bubbles, and these are created by oscillating strings as described in chapter three. Space "pinches down to a point" at a point of encounter between expanding VE-spheres because the spheres collapse, due to the encounter, and a new VE-sphere expands from the point of encounter. The QV-bubbles expanding and collapsing in the surfaces of VE-spheres are the quantum vacuum fluctuations of space. What Greene called a "flop transition" is the 90-degree rotation, or deflection, of the effects encountering each other when VE-spheres interact. The process is analogous to the deflection of equal and opposite effects that occurs when the thrusts of oscillating strings converge on a point of encounter, as described in chapter three. If it is point of encounter on which thrusts converge from all directions of space, the thrusts cancel each other at that point and are deflected into a spherical surface that is orthogonal to all dimensions of space. They become the surface of an expanding QV-bubble. *Time is created* (elapses) when QV-bubbles interact at points of encounter between the expanding surfaces of VE-spheres. The process creates the emergent reality of quantum mechanical events and also creates the changing qualia of consciousness.

String theory states that there are more than four space-time dimensions. The additional dimensions are described as "curled up" in the spatially extended dimensions. As interpreted here, this is attributable to the fact that *three* QV-bubbles are involved in an encounter between *two* VE-spheres. Two QV-bubbles interact when two VE-spheres encounter each other because the surfaces of VE-spheres consist of QV-bubbles. The third QV-bubble expands from the point of encounter. The three QV-bubbles exist simultaneously because the collapse of QV-bubbles at points of encounter and the expansion of a new QV-bubble are instantaneous ("timeless") events. In the Physics of Encounter, the "dimension" of time is a process that occurs as QV-bubbles expand and collapse. They contain the qualia of consciousness, and their interaction at interrelated points of encounter (coupled map lattices) creates the quantum mechanical reality of particles.

The four detectable dimensions of space and time are continuously re-created by the effects of VE-spheres that expand into the three dimensions of space and create time when the proliferating QV-bubbles in their surfaces interact at points of encounter. Six additional, undetectable dimensions of space are "curled up" in the spherical surfaces of the QV-bubbles that exist in the surfaces of the two VE-spheres at the instant of their interaction: three each in the two QV-bubbles that interact and collapse when the encounter occurs. The third QV-bubble expands into a new VE-sphere. As explained in chapter three, the thrusts of the oscillating strings along the three dimensions of space cancel one another at a common point of encounter. They are deflected by 90 degrees and become the surface of a new QV-bubble. The six additional dimensions are the effects of the two interacting QV-bubbles. The interaction between VE-spheres involves four detectable and six undetectable dimensions, or effects, of space and time. This is in keeping with the scenario proposed by string theorists.

In the scenario proposed here, the ten dimensions represent the sum total of effects if the dimension of time is counted only *once*, as part of a process that involves an act of observation by one observer. An act of observation, as defined by the Physics of Encounter, occurs when an encounter between two QV-bubbles, of which one contains the qualia of the observer's consciousness, creates a quantum mechanical event and a new QV-bubble that contains a new mix of qualia. Eleven dimensions are involved if we take into consideration that a co-occurrence is needed for a quantifiable event, as explained by Manthey (4/9).

Since encounters occur only in morphic resonance, the QV-bubble that contains the mind event of the observer and the QV-bubble that exerts the effect of the quantum mechanical event are always the size size. The *Planck length*, therefore, is a universal constant. It corresponds to the smallest size of the QV-bubble that exerts a measurable matter effect on the observer. The undetectable dimensions in the above scenario are *sub-Planck* dimensions, which string theorists equate with the Kaluza-Klein radius. It corresponds to the radius of the expanding QV-bubble in the surface of a VE-sphere just *before* the encounter occurs.

The expansion of a QV-bubble, as explained in chapter three, is energized by the thrusts of countless oscillating strings converging on a common point of encounter from all directions. The radius of the smallest conceivable QV-bubble is the sub-Planck length of one oscillating string. The radius of the QV-bubble exerting a *measurable* effect comprises the length of a variable number of strings, depending on the circumstances of the observation. This includes the slowing of time and the contraction of space described by relativity theories.

The hemispheres of the three QV-bubbles that co-exist at the instant of an encounter define the smallest logical segment of space, the *quaternion*. The segment of space enclosed by a quaternion is defined by two converging hemispheres and two diverging hemispheres that expand along orthogonal axes. The converging hemispheres *still exist* and the diverging hemispheres *already exist* when an encounter occurs. The diverging hemispheres are the surface of the new QV-bubble created by the encounter. The converging hemispheres are the surfaces of the two QV-bubbles expanding toward a point of encounter. The axes are orthogonal for reasons explained in chapter three. A 90-degree rotation of thrusts occurs when equal and opposite thrusts exerted along one dimension *converge* on a point of encounter. The event creates equal and opposite thrusts that *diverge* from the point of encounter.

The quaternion defines the element of uncertainty that exists when an encounter occurs. The size of the spatial segment enclosed by the four hemispheres cannot be separated from the process that creates the dimension of time. All four hemispheres cease to exist and new hemispheres are created when time begins to elapse. Heisenberg's *Uncertainty Principle* can be derived from this scenario. The proposed interpretation of a quaternion also accommodates the simultaneous creation of a negative and a positive charge predicted by *Paul Dirac*. The effect of a positive charge, as I explained, is transmitted by a hemisphere that expands away from a point of encounter, i.e. away from the observer. A negative charge is the effect of a hemisphere that expands toward the observer. The axis along which this effect impinges on the observer is orthogonal to the thrust that creates the mass, or gravitational effect, of the particle.

String theorists say that the interaction between oscillating strings creates "loops". A loop, as described by Greene, "can surround a one-dimensional piece of space like a circle".(14) The loop corresponds to the circumference of a QV-bubble expanding into a VE-sphere. As interpreted here, it plays a role in the theory of *loop quantum gravity*, which I will describe in chapter tweny. String theorists describe the loops as "one-brane" configurations. The more complex configurations, they say, consist of two branes and three branes. The branes, in this context, correspond to what I have called the effect planes that participate in the creation of elementary particles, as described in chapter twenty-one.

The string theory, according to its adherents, resolves a troublesome issue for theoretical physicists. The elementary constituents of physical reality, in that theory, are unobservable because they are strings of sub-Planck dimensionality. "The incompatibility of general relativity and quantum mechanics", Greene said, "is avoided in a universe which has a lower limit on the distances which can be accessed.(15) The Physics of Encounter proposes an enlarged interpretation of what has been called a "one-dimensional" oscillating string. The string is a line of thrust. It is a one-dimensional oscillation in the dimension of *time*. It is the force of an expansion into all three dimensions of space that is inherent in the zero-point energy (ZPE) of a quantum vacuum. The oscillation of this force is an oscillation between the explosions and implosions that are the quantum vacuum fluctuations of space. They occur in the quantum vacuum of the QV-bubbles that energize the expanding surfaces of VE-spheres. Greene has acknowledged that the string theory lacks an "overarching and systematic framework" and needs to be integrated into a "new, more geometrical framework of quantum mechanics".(16) The Physics of Encounter introduces this geometrical aspect into the scenario of interacting strings.

An *explosion*, in this scenario, occurs when the surface of a QV-bubble expands *toward* a point of encounter with the observer in the dimension of time. An *implosion* occurs when the thrusts of oscillating strings converging on a common point of encounter from all directions of space are exerted *away* from the observer. The oscillation of strings connects two types of physical reality: the hidden reality of potential energy, as defined by Ibison (12/9), and the emergent reality of measurable energy.

The hidden energy in our universe, as interpreted here, is the zero-point energy in the quantum vacuum fluctuations of space. It is released at points of encounter between the expanding surfaces of VE-spheres. The Physics of Encounter suggests that the explosive release of energy in the proverbial "big bang" that created our universe was an encounter between the hemispheres , or complementary "branes" of hidden energy. This corresponds to the big bang scenario described by the string theorist Gabriele Veneziano. Like Hawking and other astrophysicists, he argued that our universe is a "brane" and that the big bang was "a collision" between or universe and a different brane. In that type of interaction, he said, "*inflatons*" and "*dilatons*" play the decisive role. Veneziano described the inflaton as a field of potential energy that drives matter apart when the collision occurs. It counteracts the effects of gravity. The dilaton, he said, determines the strength of all interactions involving branes. It "fascinates string theorists because its value can be interpreted as the size of one of the additional dimensions of space".(17)

What Veneziano called an inflaton is an expanding VE-sphere. A dilaton is a QV-bubble in the expanding surface of the VE-sphere. QV-bubbles energize the expansion. As explained above, the dimensionality of QV-bubbles participating in an encounter between the expanding surfaces of VE-spheres contribute six additional sub-Planck dimensions to the event that creates space and time.

Like the oscillation of strings, the interaction between VE-spheres is an undetectable reality. This scenario resolves the contradiction between the conceptual constructs of general relativity and quantum mechanics. The gravitational effects described by the theory of general relativity are created at points of encounter between the surfaces of the expanding VE-spheres. The interaction also creates the quantum mechanical reality of a particle at interrelated points of encounter that include the act of observation of the particle's observer.

*NOTES*

(1) Boes (2005), conference presentation
(2) Atmanspacher *et al* (2006), pp. 3 and 6
(3) Biermann (2006), conference presentation
(4) Davies (1978)
(5) Hawking (2001), p. 188
(6) Ibid, pp. 196-197
(7) Beichler (2007), conference presentation
(8) Ibison (2006), conference presentation
(9) Ibid
(10) Einstein *et al* (1935), p. 777
(11) Bohm *et al* (1987), p. 323
(12) von Lucadou (1997), pp. 91-92
(13) Greene (2000), p. 325
(14) Ibid, p. 324
(15) Ibid, p. 164
(16) Ibid, pp. 375 and 382
(17) Veneziano (2004), pp. 32 and 35

## Chapter 20

**The hidden processes underlying current conceptual constructs (continued).**

**(a) - Photons, time, and atomic nuclei.**

This chapter proposes descriptions of the hidden processes underlying quantum gravity and loop quantum gravity. These processes, as interpreted here, are energized by the exchange of photons between the atoms of gravitationally attracted objects. I will therefore begin this chapter by summarizing how the Physics of Encounter describes the emission and absorption of photons.

The proposed scenario is based on recent research into the processes that occur inside an atomic nucleus. A process that suggests itself as instrumental in the creation and emission of photons is the *chirp wave*. As interpreted here, it contains the vacuum energy that accumulates as the QV-bubbles inside the nucleus interact. Vacuum energy is released at points of encounter between the surfaces of QV-bubbles. The evidence suggests that the wavelength of a chirp wave decreases as the energy accumulates, and that the energy of the wave erupts sporadically. The eruptions are *photons*. The energy of photons is the vacuum energy contained in the expanding surfaces of VE-spheres. The accumulating vacuum energy in a chirp wave is contained in the interacting QV-bubbles that create the quarks in an atomic nucleus. Quarks are points of encounters between QV-bubbles. Research at the Fermi National Accelerator Laboratory in Illinois (11/1) suggests that quarks have an "outer shell". As interpreted here, the outer shell corresponds to the expanding surface of a QV-bubble.

In the scenario proposed by the Physics of Encounter, photons are emitted from the nucleus of an atom and not from the electron shell. The expanding VE-surface of a photon transmits an effect in the dimension of time. The surface does not move through space (it expands in the dimension of time) and can therefore "jump over" the space occupied by the electron shell. The process is roughly comparable to the emission of bullets from the machine guns that were mounted behind the propellers of WWI airplanes. The bullet shots were synchronized so that the bullets flew through the space between the rotating propeller blades. The shots were fired at right angles to the direction of the rotation. In the same sense, time is a process that occurs at right angles to the events in the dimensions of space.

When atoms interact by exchanging photons, quarks and orbital electrons interact. The surfaces of VE-spheres that are electron orbits contain QV-bubbles that may expand into the atomic nucleus and create quarks. Conversely, some of the QV-bubbles that are quark effects in the nucleus may expand beyond the boundary of the nucleus and influence events in the electron shell. They are the effects of up quarks. These effects accumulate behind a wall of energy that is like a boundary separating the up quarks in the nucleus from the electron shell. The *asymptotic freedom* of up quarks, as interpreted here, means that these quarks cannot leave the nucleus as long as they retain their identity. Through the process of *tunneling*, the accumulated energy of up quarks escapes from the atom as the energy of photons in the *offer waves* of forward causation. The energy of photons in the *echo waves* of retrocausation is the accumulated energy of down quarks that is expelled from the nuclei of atoms by chirp waves. The expulsion of both types of photons is like the synchronized emission of bullets through the fluctuating presence of the rotating propeller blades at specific points in space, as described above.

The two types of waves, transmitting the effects of either forward causation or backward causation, originate at opposite "ends" of the atomic nucleus in the dimension of time. The nucleus, as interpreted here, is not a palpable object suspended in space. It is, rather, a configuration of explosions and implosions that involve the QV-bubbles in the atomic nucleus. Some of the details of the process through which photons are emitted as the accumulated energy of up quarks or down quarks are described in chapter eleven.

The VE-surfaces of photons created in an atomic nucleus expand instantly and become part of the quantum vacuum fluctuations of space. No time elapses between the emission of a photon and its absorption into a different atom. Time, as a process, is *created* as photons are exchanged between atoms. The events that constitute acts of observation occur in the nuclei of interacting atoms and create what I have called the gluonic light of consciousness. The qualia of consciousness are the vibrational patterns of oscillating strings in the quantum vacuum of interacting QV-bubbles. Gluons are the subatomic equivalent of photons. Gluons connect the quarks in atomic nuclei. Photons connect interacting atoms. The interacting atoms in a brain exchange *biophotons*.

## (b) - Quantum gravity and dark matter. The absorption of photons and sub-Planck shifts in the location of point-particles.

The process of gravitational attraction, as interpreted by the Physics of Encounter, can be equated with the concept of quantum gravity because it involves the same type of unobservable interactions that create quantum mechanical events. In this scenario, the gravitational field consists of expanding VE-spheres interacting at points of encounter. The scenario suggests that the quarks and electrons in an atom are continuously re-created at points of encounter between the surfaces of VE-spheres. Under the influence of gravity, all of these re-created points within an atom are minimally displaced into the same direction, so that the whole atom moves into that direction.

Initially, both quarks and electrons are dimensionless point-particles, but the points expand into QV-bubbles, and then into VE-spheres. The VE-spheres causing a gravitational displacement originate *outside* the atoms of the object that changes its location due to the influence of a gravitational field. The encounters between the expanding VE-surfaces occur *inside* the atom that is influenced by quantum gravity. Each encounter is an encounter between the oscillating strings that energize the expansion of the VE-surfaces. Each point of encounter, therefore, is the endpoint of oscillating strings expanding toward a common point of encounter. This creates a new QV-bubble, as described in chapter three. The successive displacements caused by quantum gravity correspond to the unobservable sub-Planck length of the oscillating strings that create the QV-bubbles in the surfaces of the interacting VE-spheres.

If the quarks and electrons in an atom are continuously and simultaneously re-created with sub-Planck shifts in location, and if the location is shifted into the same direction, the entire atom moves. The interactions inside the atoms of an object occur as the atoms absorb photons, which then disintegrate into the effects that cause the quantum jumps of electrons and create the quark effects in the atomic nucleus. The "orbits" of the electrons are the expanding and collapsing surfaces of VE-spheres. The "wavelength" of the electrons is the radius of the QV-bubble that is created at the point of encounter.

When the location of an orbital electron is changed through quantum gravity, the location changes *within* the orbit. That is not a quantum jump. Orbital electrons constantly change their location in an orbit because they are constantly re-created at points of encounter between the surfaces of VE-spheres. In that sense, the electrons "orbit" or circle around the atomic nucleus.

The process of "gravitational attraction" occurs when the motion of a small object is influenced by the effect of the quarks in a large object. Quarks exert gravitational effects as they expand into VE-spheres. The quark effects cause the small object to move toward the large object. Since quark effects occur in both objects, the large object also moves, but the motion is minimal if the two objects are of very different size. To simplify the description of the process, I will describe only the motion of the small object relative to the large object. It could be a small object "falling" to the surface of the earth.

To visualize the process, it is important to understand that the process of "falling" can be like a "horizontal" motion in outer space, i.e. a motion along the line between the two objects that influence each other. Many other objects may be located nearby, but their influence is of the same kind and will be kept out of the picture presented here. Of importance in this simplified description of quantum gravity is the location of the quarks and electrons in the atoms of the small object with reference to the large object.

To describe how the quark effects in the atoms of a large object influence the motion of atoms in a small object, I will use the terms "left" and "right" to describe the location of the small and the large object relative to each other. If a large object is located to the left of a small object and influences the small object through quantum gravity, the re-creation of orbital electrons in the atoms of the small object is more likely to occur to the left of the atomic nucleus, because the re-creation occurs where the expanding surfaces of VE-spheres encounter each other. This, in turn, influences the re-creation of quarks inside the atomic nucleus. The encounters that create those quarks are, therefore, more likely to occur on the left side of the particles in the atomic nucleus.

The entirety of all encounters between the surfaces of VE-spheres re-creates the atoms themselves. In the world of quantum mechanics, as interpreted here, the continuous motion of particles and atoms is their continuous re-creation at changed locations defined by the points of encounter between the surfaces of VE-spheres. In the scenario described here, all atoms of the small object move to the left, toward the location of the large object.

Quantum gravity, in spite of its name, causes a smooth, point-by-point sub-Planck displacement, or motion, of the atoms in a physical object. It does not involve any quantum jumps. The quark effects are transmitted by the surface of photons that impact on an atom from all directions, but the largest number of such effects originate in the large object that influences the motion of the small object.

The process illustrates the paradoxical, or dialectical, nature of the unobservable influences that create the quantum mechanical reality of observable events. What is called gravitational "attraction" is caused by the *impact* of an unobservable effect on an observable reality. The impact comes from a direction that is opposite to the direction of the motion caused by the impact.

The above scenario explains the *acceleration* of a small object that is "falling" toward a large object. The distance between the two objects determines the probability of the encounters that cause the gravitational "attraction". The encounters are random events. They occur only in morphic resonance. When the two objects are still far apart, relatively few of the quark effects (QV-bubbles) in the expanding surfaces of the VE-spheres that originate in the large object will impact on the atoms of the small object. The quark effects do not reach the small object because the photons that transmit them through space (the expanding VE-spheres) "disintegrate" along the way. The surfaces of the VE-spheres that act as photons contain unobservable vacuum energy, or potential energy as defined by Ibison (12/9), until they are encountered, or absorbed into an atom. In the requisite configuration of encounters, the absorption is an act of observation. For reasons explained below, photons created where atoms exist may also disintegrate when they traverse so-called "empty" space, where no acts of observation occur. Space devoid of observers contains no atoms. As Wheeler said, we contribute to the ongoing creation of the universe we observe. Wheeler's startling argument was headlined in the journal *Discover* (7/10) by an appropriately radical question: "Does the universe exist if we're not looking?"

To look at something, we need the light that is transmitted through space by the photons exchanged between interacting atoms. A photon disintegrates when it is absorbed into an atom. Its surface "splinters" into the QV-bubbles that are contained in the surface of a VE-sphere. Its energy impacts on the surfaces of the VE-spheres that are the "orbits" of electrons. Some of the QV-bubbles expand into the atomic nucleus and generate quark events. The probability is high that a photon will disintegrate as it travels from the large object to the small object if the two objects are still far apart. The probability is reduced as the distance between the two objects decreases. The number of transmitted quark effects that cause the gravitational attraction increases accordingly. A higher number of quark effects impinging on the atoms of the small object from a location to the left of the object speeds up its motion toward the left.

An atom consists of QV-bubbles in the surfaces of VE-spheres that are interacting in a stable configuration. The surface of a photon disintegrates into its constituent QV-bubbles when the photon is absorbed into an atom. Inside the atom, all of the QV-bubbles are encountered in morphic resonance. The photon "fits" into the atom. This means not only that the photon (the VE-sphere) has to have a certain size but also that its surface has to contain the requisite wave pattern to "enmesh" with the wave patterns of the atom.

A process analogous to the absorption of a photon into an atom occurs in the vacuum of outer space, where there are no atoms, only quantum vacuum fluctuations. Invisible configurations of QV-bubbles exist as quantum vacuum fluctuations throughout space. Photons are VE-spheres that have expanded from QV-bubbles in the surfaces of VE-spheres. They are, in other words, VE-spheres that exist as superpositions in the surfaces of larger VE-spheres. Photons travel through space in the expanding surfaces of larger VE-spheres. Even if there are no atoms that absorb the photons, the photons will eventually disintegrate because all of space is pervaded by the quantum vacuum fluctuations of interacting VE-spheres, even the vacuum of so-called "empty" space. When the layers of expanding and collapsing QV-bubbles that are the surfaces of photons are encountered, in morphic resonance, by the surfaces of other VE-spheres in the fluctuating vacuum of "empty" space, they collapse and the photon ceases to exist. This happens to some of the photons that originate here on earth, or anywhere else where there are atoms and solid objects, and then travel into the emptiness of outer space. The disintegrating photon deposits its gravity effects into the space where there are no atoms. This accounts for the *dark matter* in the distant regions of our universe.

In the vacuum that exists in the outer regions of the universe, there are no atoms emitting and absorbing photons. The prevailing view is that photons travel, invisibly, through the "empty" regions of outer space. Observers traveling through that space would look into total darkness. The invisible photons, in scenario proposed here, are the same physical reality as the photons that are exchanged between atoms. They are the vacuum energy of QV-bubbles in the interacting surfaces of VE-spheres. They are part of the quantum vacuum fluctuations of space.

The above scenario is consistent with the theory of general relativity. The potential energy of a gravity effect exists at every point in the universe, in the darkness of outer space and in the light that floods our atmosphere when photons are exchanged between interacting atoms.

## (c) - Loop quantum gravity. Wheeler's cosmic feedback loop. Mini-loops and "bit bangs". Angular momentum and centrifugal forces.

The process that creates quantum gravity effects needs to be understood in the context of loop quantum gravity. The loop is a *time loop* within our universe. John Wheeler, according to Folger (7/10), described our universe as an "enormous feedback loop (...) in which we contribute to the ongoing creation of not just the present and the future, but to the past as well". The word "we", in this context, refers to all acts of observation by all of mankind, past, present, and future.

Acts of observation occur as photons are emitted from atoms and absorbed into the atoms of brains. The mind events that are created by these processes exert an influence on the physical reality outside the brain. This influence is transmitted by the expanding surfaces of VE-spheres. The existence of a time loop in our universe means that the effect of every event will eventually return to its point of origin. As defined here, an *event* is an encounter between the expanding surfaces of VE-spheres. The *effect* of the event is an expanding VE-surface. It consists of hemispheres expanding into opposite directions of time, relative to the point of encounter. Every event, therefore, has two effects. Within the time loop, both of these effects will eventually encounter each other again and create another event.

To visualize the process, imagine that the point of encounter where an event occurs is a spot on the rim of a rotating, vertical disk. If you face the disk and the spot is on the far left side of the rim, the spot moves from left to right in an upper semicircle when the disk rotates clockwise. The rotating rim is a loop. The spot reverses its motion along the left-right axis when it has moved as far to the right as possible within the limits of the two-dimensional loop. The spot will then swing from right to left in a lower semicircle.

The two endpoints of the left-right axis connecting the complementary half-loops correspond to the "big bang" and the "big crunch" in the four-dimensional loop that is our universe. The universe pulsates within the loop. It explodes into pure energy in a big bang and acquires mass. At some point, the anti-gravitational effect of the explosion is overcome by the gravitational effect of the accumulating mass and the universe once again collapses into an infinitely dense point, in endless cycles. The conceptual construct of rotation within a four-dimensional loop may seem far-fetched, but it has been used and painstakingly elucidated by some of the keenest minds in theoretical physics, in mathematics and philosophy.

Does the universe *as a whole* expand and collapse? Or does it just rotate, relative to the observer, exposing the observer to a continuously changing kaleidoscope of effects at its boundary? The following information was taken from a paper published by the physicist Wilfried Kugel.(1) Kurt Gödel, who worked with Einstein at Princeton University, argued that our universe is a rotating system of coordinates, and that every observer of the universe is at rest relative to that coordinate system. A similar argument was made by Hermann Weyl. He had worked with Einstein at the Technical University in Zurich. Einstein, too, used the concept of time loops in a rotating coordinate system. Hermann Minkowski, Max Born, Hans Reichenbach, and Brandon Carter presented analogous conceptual models. An international conference on the issue of cosmic rotation was held at the Technical University of Berlin in 1998.

The kaleidoscope of effects that presents itself to the observer of the universe includes the phenomena of black holes at various locations in the outer regions of cosmic space. Black holes are enormously powerful concentrations of gravitational forces. Light waves, which are only "bent", or deflected, in the vicinity of moderate gravitational forces, are "swallowed" in their entirety by black holes. Theoretical physicists have conjectured that the light disappears into a "different universe".

Kugel has argued that our universe could be metaphorically described as a rotating black hole. He pointed out that, even before the relativistic interpretations of time and space, Isaac Newton had proposed a model of a rotating universe. Newton had recognized the correspondence between the effect of gravity and the centrifugal force of rotation. The arguments presented by Minkowski and Born, Kugel said, are based on the same insight. If our universe rotates in a four-dimensional continuum, the rotation would manifest itself as an acceleration, or centrifugal force, in the three spatial dimensions of our universe. This would account for the red shift of the light reaching us from the distant regions of our universe. The red shift of the spectral lines has been interpreted as evidence that the universe is expanding and was created by an explosive force (the "big bang") that continues to push the matter in our universe away from its observers. But it could also serve as evidence, Kugel argued, that our universe is rotating and does not owe its existence to any singular event like the big bang as the one and only event that created all reality.

The Schwarzschild radius that describes the enormous size of the gravitational effect attributable to a cosmic black hole, according to Kugel, is identical with the gravitational radius of the entire mass in our universe, i.e. with Einstein's "world radius". In that sense, Kugel explained, our universe is like a black hole.

Gabriele Veneziano has suggested (5/1) that the center of a black hole is not a point in space, but a point in time. In the cosmological model proposed by Lee Smolin (5/2), our universe continuously gives birth to additional universes through black holes. Our own universe, according to Veneziano (5/3), was also created inside a black hole.

Admittedly, the argument that we live in a black hole has a strange ring to it. The same goes for Hawking's metaphor that "we live on a brane" and the other counter-intuitive scenarios described in the scientific literature about the mysteries of our universe and our existence in it. Many questions are unresolved. In 2009, the American Astronomical Society took note of the fact that the Milky Way, the galaxy in which our earth is located, is far more massive than previously calculated. As reported by Kenneth Chang in the *International Herald Tribune*, the motion of the sun around the center of the galaxy is about 15 percent faster than had been calculated, which means "that the galaxy must have more mass - about 50 percent more - to generate a stronger gravitational pull to keep hold of the Sun, as well as all its other stars. (...) Determining the shape, size and mass of the Milky Way", the article continued, "is difficult. Most of the mass is in the form of invisible dark matter, a component that far outweighs the ordinary matter in stars and gas clouds."(2)

A scenario for the creation of invisible dark matter is proposed in section (b) of this chapter, based on the disintegration of the expanding VE-surfaces of photons that occurs when light travels into the quantum vacuum fluctuations of "empty" space. The cosmic scale of events is mirrored in the quantum mechanical reality that is created by interacting VE-spheres. Loop quantum gravity comes into play when effects are propagated as forward momentum and as *angular momentum* by the expanding surface of a VE-sphere. The angular momentum is propagated within the surface into both directions of time by the thrusts of oscillating strings.

As summarized in the introduction to chapter nineteen, strings oscillate in all three dimensions of space and create the dimension of time at a common point of encounter. When two VE-spheres encounter each other, their circumferences and the point of encounter are located in a common plane. I have called it an *effect plane*. At the point of encounter, the strings oscillate along two axes that are transversal and lateral to the circumference, and along one axis that is longitudinal within it. The longitudinal oscillation is propagated within the circumference as angular momentum.

The oscillations within VE-surfaces, propagated into both directions of time, are the unobservable potential energy in the offer waves and the echo waves of electromagnetic radiation. The potential energy is the undetectable vacuum energy in the expanding surfaces of VE-spheres. It becomes measurable energy when it is absorbed into atoms. The measurable energy is the energy of photons. The two types of wave permeate each other as they expand forward in time and backward in time, respectively, creating spin effects as they travel toward the atoms that will absorb the photons. I described the creation of spin in chapter twelve.

In this scenario, every potential gravity effect that is transmitted between atoms in the expanding surface of a VE-sphere will eventually return to its point of origin in a time loop that encompasses the entire universe. The cosmic loop of our universe consists of infinitely many mini-loops. They contain the points of encounter between the surfaces of VE-spheres that create the "bit bangs" described by Michael Manthey (4/9). The hemispheres that create bit bangs at points of encounter collapse. But with every new VE-sphere created by a bit bang, another mini-loop of expansion and collapse begins. The potential energy transmitted by the hemispheres expanding into the depths of space corresponds to the effect that the journal *Science*, according to Klein (4/10), called "cosmic anti-gravitation". Many of these hemispheres are encountered in morphic resonance by hemispheres expanding into the opposite direction. Both of the hemispheres encountering each other collapse, but others continue to expand until they have gone "full circle" within the space-time continuum of our universe. What happens at the endpoints of the cosmic diameter is open to debate. Is our universe pulsating like a heart, in endless cycles of expansion and collapse? Or is it one of many "parallel" universes, hidden in the black holes that are created at points of encounter between invisible cosmic hemispheres?

Both scenarios correspond to the physical processes that underlie the interaction between mind and matter. The convergence and divergence of hemispheres create the quark events in atomic nuclei. Where hemispheres converge on a point of encounter, a down quark increases the density of elementary matter events. Where hemispheres diverge from a point of encounter, an up quark decreases the density of elementary matter events. Gluons interconnect the quark events in the nucleus of an atom. Quark events create the photons that interconnect atoms. The qualia of consciousness exist within the gluonic light of interconnected atoms. Each conscious mind is like a separate universe.

*NOTES*
(1) Kugel (1996), pp. 224-236
(2) Chang (2009), p. 2

# Chapter 21

The hidden processes underlying current conceptual constructs (continued).

**Particle classification. Inertial mass and gravitational mass. The electric charge and the effect planes of photons. The Lorentz invariance of moving particles. Particle decay.**

In the classification of particles, one of the main distinctions is between the three *families* of particles and the three *generations* of particles within each family. The criterion for distinguishing between the three families of particles is the fact that the expanding surface of a VE-sphere may be simultaneously encountered at one, two, or three different points. In the following description of the process, I will identify that VE-sphere as sphere A and its surface as surface A. Three encounters occur on surface A if three VE-spheres expand toward it along the three dimensions of space. When the encounters cause the collapse of sphere A, three new, interrelated VE-spheres expand from the three points of encounter. In a coupled map lattice, this is the configuration of the particles called *baryons*. They are the heaviest and most complex particles in that family of particles. If only two VE-spheres expand toward separate points of encounter with surface A, along two dimensions of space, the configuration created by that interaction is a *meson*. The simplest particle in this family of particles is a *lepton*. It is created when the surface of sphere A is encountered by one other VE-sphere.

To describe the three *generations* of particles, I will use the concept of *effect planes* that I mentioned in chapter twenty, section (c). They correspond to the three orthogonal planes within which you can cut through an apple, for example. They are the frontal, transversal, and sagittal planes. An effect plane is the plane within which an encounter between two VE-spheres occurs. The effect plane contains the thrusts exerted toward each other by the expanding circumferences of interacting VE-spheres. The line that connects the points of origin of the interacting VE-spheres is located within that plane. I will call the line an *axis of encounter*. It corresponds to one of the dimensions of space. If two additional VE-spheres expand toward the same point of encounter along an orthogonal axis, the process involves two effect planes. The interaction involves three effect planes if a total of six VE-spheres expand toward a common point of encounter along three orthogonal axes of encounter. The three axes correspond to the three dimensions of space.

The criterion for distinguishing the three *generations* of particles is the number of effect planes that contribute to the thrust of their expanding surfaces. In visualizing this scenario, it is important to keep in mind that the surface is a hidden reality. It consists of the infinitely many points where an encounter may occur. The points are interconnected by oscillating strings. The spherical surface may be encountered at different points in any of the three dimensions of space. It may be encountered at one, two, or three points simultaneously, as I mentioned above in describing the criterion for distinguishing the three *families* of particles.

In this scenario, the effect planes are not static entities. *Every particle is a wave* of collapsing and expanding surfaces. The surfaces, in turn, consist of waves that are layers of expanding and collapsing QV-bubbles, i.e. superpositions. These provide additional criteria for distinguishing the characteristics of particles, but I will not describe them here.

For every particle, there is a corresponding *anti-particle*, because the surfaces of particles may expand either toward or away from the observer in the dimension of time. Positrons, for example, are the anti-particles of electrons. As Feynman said, positrons are particles that "travel backward in time".

Electrons and positrons belong to the family of leptons. The second generation of leptons are called muons. The second generation of mesons are called pions. The second generation of baryons are called lambda particles. The third generation of leptons, mesons, and baryons are called tau, kaon (or upsilon), and sigma, respectively.

The above scenario also accounts for the particles called *quarks*. They are created at points of encounter inside mesons and baryons. These two families of particles are collectively called *hadrons*. I have already mentioned that quarks, in this scenario, expand into QV-bubbles and therefore have a surface. This is based on research with particle accelerators suggesting (11/1) that quarks have "an outer shell". The *flavor* of a quark corresponds to the number of effect planes that contribute to the thrust of its surface. The *color* of a quark depends on the orientation of the effect plane in the three dimensions of space, relative to the observer of the particle that contains the quark.

In this scenario, the expanding surfaces of quarks encounter one another inside the hadrons that contain the quarks. New quarks are created at the points of encounter. The mosaic of encounters creates quarks at various locations inside the particle that contain them. If the location is near the boundary of the particle, the expanding surface of the quark may become one of the surfaces of the observed particle (a meson or baryon). This is why the characteristics of quarks can be deduced from the characteristics of the particles that contain them. The quarks themselves are unobservable.

The Physics of Encounter distinguishes between quarks and quark effects. A quark is a point of encounter inside the particle that contains it. A quark effect is transmitted by the expanding surface of the QV-bubble that is created at the point of encounter. The surface consists of hemispheres, of which one expands toward the observer of the hadron, i.e. toward the boundary of the hadron, while the other hemisphere expands away from the observer, "deeper" into the hadron. The encounters and the expansion of the hemispheres occur along the dimension of time. An expansion toward the observer is the effect of an *up quark*. The expansion away from the observer is the effect of a *down quark*. The effects of quarks on the mass of particles are explained in chapter eleven.

The above scenario provides a plausible basis for classifying particles because it explains a large variety of particle characteristics. It explains, for example, why the *inertial mass* of a particle is quantized, whereas the *gravitational mass* of a particle is not. The motion of particles due to the influence of gravity is based on the process that I described in section b of this chapter with reference to the gravitationally influenced motion of atoms. The effects of "quantum gravity" cause sub-Planck, point-by-point displacements. The same applies to the points of encounter that establish the identity of a particle. The configuration of interrelated points is continuously re-created within the gravitational field. This is the basis for the *Lorentz invariance* with respect to the motion of particles that are not accelerated by an additional source of energy. In the case of accelerated motion caused by influences other than gravity effects, the resistance to change (inertia) is rooted in the configuration as whole. The configuration establishes the quantized mass of the particle. As Lederman explained (10/6), mass "is not an intrinsic property of particles", but is acquired in a Higgs field through "system configuration".

The "systems", in the scenario proposed here, are expanding VE-spheres that interact when their surfaces encounter one another in morphic resonance. In this interaction, VE-spheres collapse when they are encountered and new VE-spheres expand from the points of encounter. This is the process through which new particles are "discovered" in particle accelerators. It is not so much a discovery as it is an artificial creation of particles that do not normally exist on the surface of the earth. The "collisions" of accelerated particles are encounters between the expanding surfaces of VE-spheres that would not normally occur.

The process that occurs in a particle accelerator is like the unobservable process in a hypothetical spaceship that accelerates to a speed close to the speed of light. The "system configurations" that constitute the mass of the spaceship, i.e. the atoms and their constituent particles, are reduced in size, due to an increased frequency of encounters with the interacting VE-spheres that are the quantum vacuum fluctuations of space. As the spaceship travels through those fluctuations, the higher density of encounters causes the relativistic contraction of the space occupied by the accelerated object.

In a particle accelerator, a higher density of encounters occurs because the particles are accelerated into opposite directions and encounter one another in "collisions" that are interactions between the surfaces of VE-spheres expanding toward one another. In contrast to the process that occurs in an accelerating spaceship, two system configurations interact. Both have already been compacted by a process analogous to the one that occurs in the spaceship. In this unique situation, the surfaces of many extremely small VE-spheres encounter one another. Since encounters are random events that occur only in morphic resonance, the high frequency of encounters increases the probability that the surfaces will be encountered at more than one point. Through the process described above, encounters at two or three points create additional effect planes. Particles comprising more than one effect plane are higher-generation particles with more energetic surfaces. Under ordinary circumstances, such extremely small, highly energetic particles would not be created.

According to conventional wisdom, particles are smashed when they collide in a particle accelerator. A more appropriate description is that the particles are not re-created because the normal encounters that re-create the system configuration of those particles do not occur. Instead, additional encounters occur that create additional, higher-generation particles.

The scenario of interacting VE-spheres proposed by the Physics of Encounter also explains the process that occurs when particles decay. The surfaces of the particles that contain quarks are, as described above, the expanding quark surfaces that have not been encountered inside the particle. When these quark surfaces become the boundaries, or surfaces, of particles, they will continue to expand unless the particle that contains the quarks is observed. The act of observation occurs when the surface of a VE-sphere that is a photon impacts on the surface of the particle. This encounter between two VE-surfaces creates a new VE-sphere. The two hemispheres of that VE-sphere expand into opposite directions of time, relative to the observer of the particle. One hemisphere expands into the particle and the other hemisphere expands toward the observer. When the hemisphere that expands into the particle is encountered, another quark event is created inside the particle. The hemisphere that expands toward the observer is, for one instant only, the re-created surface of the particle. When the expanding surface reaches the observer, this is the event through which the particle is observed.

The act of observation re-creates the particle because it creates another quark event inside the particle. This event re-creates the quark effect that enabled the observer to become aware of the particle. A particle containing quarks is kept in existence by continuous acts of observation. As explained in chapter nineteen, the observer does not have to be physically present at the location of the particle. At any specific instant, the observer's mind event is contained in one of the layers of the QV-bubbles that energize the expanding surface of a VE-sphere. The surface is a probability wave. The layers of QV-bubbles are superpositions in a probability wave. Each QV-bubble in a layer of QV-bubbles contains, at a specific instant, identical qualia of the observer's consciousness. The *lifetimes* of particles vary because the probability of the requisite encounters varies with the complexity and other characteristics of the system configurations that characterize the particle.

A particle *decays* if it is not re-created by acts of observation. The quark surface that continues to expand beyond the boundary of the particle will continue to expand until it is encountered by an expanding VE-surface that does not contain the mind event of the particle's observer. Such encounters create less complex particles, which are then observed as products of the decay.

This scenario explains why *mesons are unstable* system configurations. In contrast to baryons, which are three interrelated VE-spheres, mesons consist of only two VE-spheres. Baryons contain three quarks, mesons contain only two. Mesons can only be observed along two axes of encounter, not along the three axes that correspond to the three dimensions of space. Mesons have, so to speak, an "open flank" through which the surfaces of VE-spheres can expand directly *into* the meson, instead of encountering the *surface* of the meson. When the surface of a VE-sphere expands into the meson and is encountered there, the encounter creates one of the two quark events that characterize a meson. The quark surface can expand through the "open flank" into the ambient space. When the surface is encountered there, the particle created by the encounter is observed as a lepton. The meson has decayed into the lepton and the other quark that was inside the meson. The remaining quark loses its identity as a quark, because there are no one-quark particles. The equivalent of a quark in the simplest family and lowest generation of observable particles, the leptons, is the point of encounter between two VE-spheres. Quarks, the unobservable particles inside the particles called hadrons (mesons and baryons), are categorized as *fermions*.

A *photon*, like a quark, does not have a spatial extent as a particle. When a photon impacts on a surface, it exerts an effect at a *point*. As part of a light wave, a photon is the expanding surface of a VE-sphere. The criterion that distinguishes the surface of a photon from the surface of other VE-spheres is the number of effect planes that energize the expanding VE-surface. In contrast to the observable particles that have a spatial dimensionality and are energized by up to three effect planes, a photon is energized by *four effect* planes. Three effect planes exist in the three spatial dimensions of the atom that *emits* the photon. The fourth effect plane is located in the surface of the atom that *absorbs* the photon. It is the Euclidean segment of that curved surface and contains the point of encounter. The four effect planes of photons connect spatially separated atoms in the dimension of time when these atoms exchange photons. No time elapses between the emission of a photon and its absorption into a different atom because the expansion of a VE-sphere occurs instantly.

When a photon impinges on the surface of an atom, it triggers a process that is analogous to the process that occurs on the surface of a particle during an act of observation, as described above. When the surface of the atom and the surface of the photon encounter each other, a new VE-sphere is created at the point of encounter. One hemisphere of that VE-sphere expands into the electron shell of the atom, and the other hemisphere expands away from the atom, toward the observer of the atom. When the hemisphere that expands away from the atom is encountered in an act of observation, the particle created by that encounter is a negatively charged lepton: an electron. This explains why electrons are observed where atoms absorb photons.

The scenario also provides a basis for describing the process that creates an electric charge. If it is a *negative charge,* its effect is observed (measured) at a point of encounter within the configuration of encounters that establishes the spatial dimensionality of the particle. One of these points is the location of the observer, i.e. a point of encounter between VE-surfaces, of which one contains the qualia of the observer's consciousness. It is the point where an act of observation occurs when the particle exerts an effect on the observer. The effect of a negative charge is exerted at that point if the surfaces of two additional VE-spheres converge on the same point of encounter along an orthogonal axis, within the dimension of time. Since these surfaces, too, expand *toward* the observer, their encounter creates an additional effect on the observer of the particle. In contrast to the *configuration* of encounters that establishes the dimensional extent of the particle, the effect of an electric charge is like a pinprick at a point in space and time.

The effect of a *positive charge* occurs within the same system configuration that establishes the spatial dimensions of the particle, but the effect of the positive charge is directed *away* from the observer within the dimension of time. The effect is exerted by the hemispheres that expand away from the point where the encounter has created a negative charge and a new VE-sphere.

The *fractional charges of quarks* can be geometrically derived from the configuration of interacting VE-spheres that create the three quarks in protons and in neutrons. Protons have a positive charge. They contain two up quarks and one down quark. Neutrons are electrically neutral. They contain two down quarks and one up quark. For brevity, the quark configurations of protons and neutrons are called u-u-d and d-d-u, respectively. A free neutron decays into a proton and two additional particles: an electron and an anti-neutrino. Anti-neutrinos and neutrinos have no electric charge. The charge of an electron is, by definition, equated with a full-integer value of minus one. In the quark configurations of protons and neutrons, u-quarks have a charge of +2/3 and d-quarks have a charge of -1/3. For protons, these values add up to +1. For neutrons, they add up to zero.

As interpreted here, d-quarks are located where hemispheres *converge* on a point of encounter, and u-quarks are located where hemispheres begin to *diverge* from a point of encounter. The creation of the three quarks in the lowest generation of protons and neutrons, therefore, involves six interacting surfaces. They are the hemispheres of VE-spheres that expand and collapse at points of encounter. The effects of quarks are transmitted by these surfaces. In a process that is too complex to be described here, four of the VE-surfaces expand toward a common point of encounter when a neutron decays. One of the four VE-surfaces contains the mind event of the observer. The interaction occurs along two orthogonal axes of encounter. It results in the observation of an electron. The process that results in the simultaneous creation of a particle and its electric charge, as described above.

The interaction changes the neutron into a proton because the *convergence* of two hemispheres that creates the negative charge is instantly followed by the *divergence* of two hemispheres from the point of encounter. They exert the effect of a positive charge. Together, they are the expanding surface of the new VE-sphere that is created at the point of encounter.

The anti-neutrino is not observed at the location where the decay occurs. The particle is like an incomplete, virtual particle. It consists of either one of the two *unencountered* hemispheres that expand into opposite directions of time, away from the location where the decay occurred. The virtual particle becomes an actual particle when either hemisphere is encountered. The other hemisphere, which is not encountered, represents the residual energy that is released by the process of the neutron decay. It is "not accounted for" in terms of particles. The existence of neutrinos has been confirmed through acts of observation at locations that are not identical with the site of the neutron decay. Since the unencountered hemispheres that expand away from the site of the decay expand into opposite directions of time, the observers of the elusive neutrino effect are either past observers or future observers relative the time and the site of the neutron decay.

The scenario explains why protons apparently have an unlimited lifetime while they exist as free particles outside the nuclei of atoms and if they are not destroyed in particle accelerators. Particles containing quarks do not decay while their multiple surfaces are re-created by acts of observation. The positive charge of a proton indicates that the proton is continuously being observed, because a positive charge is an effect that is created by an act of observation. It is the effect of a VE-surface that expands away from the observer in the dimension of time.

Here, again, it is important to keep in mind that an observer does not have to be physically present at the location of the particle. The requisite influence of the observer is contained in the expanding surface of a VE-sphere, which is a probability wave. The wave expands into both directions of time. This means that the acts of observation that keep a particle in existence could be the effects of mind events that have already occurred or will occur in the future. Encounters that re-create the surfaces of particles at points of encounter are random events. Encounters occur only in morphic resonance.

Of equal importance, however, is the fact that "an observer" is not just a point on a probability wave. As a living physical reality, an observer exists only where atoms exist. At any specific instant, the observer's mind event is created in the nuclei of interacting atoms. A mind event created by an act of observation is an event within the quantum vacuum of a QV-bubble that creates the qualia of the observer's consciousness. Protons are re-created because the universe we are observing contains atoms, and because interacting atoms create human mind events that energize probability waves.

The same is true for all aspects of the physical reality in which we exist. As observers, we are, as Wheeler said (7/10), contributing to the ongoing creation of our universe. This scenario creates the entire spectrum of elementary physical phenomena as defined by the current conceptual constructs. The final chapter of this book proposes some experiments to test the validity of the proposed scenario.

# Chapter 22

**Proposals for experiments. The CERN experiment and Lederman's "God Particle". Metaphors, science, and consciousness-related influences.**

Physicists rightfully insist that physical theories should be verifiable by experiments. The Physics of Encounter, as explained at the beginning of this book, is not a theory. It proposes a metaphorical model of mind/matter interactions to stimulate research. It points the way to a theory. Experiments confirming the validity of the proposed geometric model of mind/matter interaction will have to show that the effect of ambient quantum vacuum fluctuations on a detector system can be treated as a mind-related variable correlated with detectable changes in random quantum events. The experiments will test the hypothesis that the effect is changed by a higher incidence of mind events in the vicinity of the detector system. The effect should occur irrespective of the intentionality or cognitive content of the mind events, although such factors might influence the size of the effect.

The predicted change in the effect is based on the assumption that the influence of mind events can be equated with quantum mechanical probability waves. In the proposed metaphorical model these waves are what I have called the expanding surfaces of VE-spheres interacting at points of encounter. The expansion and collapse of these surfaces are quantum vacuum fluctuations. The surfaces consist of what I have called QV-bubbles.

A detector system that could be used for the proposed experiments is a cloud chamber, or bubble chamber. The variable to be introduced into the experimental setting is the presence or absence of a large number of humans, or possibly animals. When they are assembled near the cloud chamber, they could be subjected to a variety of sensory stimuli. The high number of mind events is assumed to result in a high number of probability waves expanding into the cloud chamber. The Physics of Encounter predicts that these waves will influence the quantum events that are detected in the cloud chamber.

A cloud chamber contains vapor in which electrically charged particles leave traces because the electric charge causes a localized condensation of the vapor. Based on the proposed model of mind/matter interaction, the Physics of Encounter predicts that the introduction of additional mind events into the experimental setting will produce the following detectable results:

1) A higher incidence of random *particle materialization*. The prediction is based on the hypothesis that particles are created by co-occurrences of random encounters between the expanding surfaces of VE-spheres. A higher number of encounters should result in a higher number of co-occurrences in coupled map lattices.

2) A decrease in the average *curvature of trajectories* of particles traversing the cloud chamber. The prediction is based on the hypothesized scenario of *quantum gravity* described in chapter twenty. It suggests that the motion of a particle is a succession of sub-Planck displacements of the particle at points of encounter between VE-surfaces. A higher incidence of encounters increases the velocity-related mass of the particle. This decreases the vector components of the magnetic effects exerted on electrically charged particles traveling through the cloud chamber.

3) An inverse relationship between the number of observed effects and the *distance* of the assembled participants from the cloud chamber. This also follows from the hypothesized scenario of quantum gravity. The probability of random encounters inside the cloud chamber is reduced if these encounters are pre-empted by encounters in the space that separates the participants from the cloud chamber. VE-spheres collapse when their surface is encountered in morphic resonance. A large separation increases the probability that the point of encounter will lie outside the cloud chamber.

4) An *increased incidence of meson decay*. The prediction is based on the classification of particles and the scenario of particle decay proposed in chapter twenty-one. Since mesons are unstable two-quark particles, one of the three dimensions of space is an axis of encounter that constitutes the "open flank" of a meson. An increase in the number of VE-surfaces expanding into the cloud chamber increases the probability that one of the VE-surfaces will expand into the meson along that axis of encounter and cause the decay of the meson.

An experimental setting to verify the above predictions with a sufficiently large number of human participants will be difficult to establish because the participants could only be intermittently present for relatively short periods of time. An initial experiment could determine whether the predicted effect of mind events is sufficiently large when animals are used instead of human participants. The assumption is that animals, too, experience mind events. In an initial experiment, therefore, a large number of animals could be put in cages in close proximity to a cloud chamber.

If the initial experiment confirms that the predicted effects occur, follow-up experiments could determine whether the assumption is correct that the effects are enhanced if the animals are subjected to additional sensory stimuli. This is, of course, also an option if the predicted effects fail to occur, but might be detectable if the effect is enhanced. The purpose of the following proposals is to point out the spectrum of possibilities, not to suggest that they all be implemented.

In addition to alternating visual and auditory stimuli, attention-grabbing events could be produced by installing moving objects in the cages housing the animals. Random stop-and-go movements would assure occasional tactile sensations. Food could be put into the cages at irregular intervals. A variety of smells, some noxious, some pleasant, could be introduced in alternation. The sensory input should fluctuate in quality and intensity to preclude monotony.

If the experiments with animals produce the predicted results, more follow-up experiments with human participants could be envisaged. That would also be an option if the experiments with animals fail to yield the predicted results. The mind events of humans might produce a more readily detectable effect. If the mind events of animals as well as humans create detectable effects, it might be of interest to determine whether the strength and, possibly, the type of effects differ with respect to the mind events of various animal species and of humans.

The experiments with human participants could investigate the relative strength of the effects generated by variables such as (a) cognitive efforts, (b) emotional states, (c) sensory input. This is of interest because the Physics of Encounter suggests that the qualia of consciousness are created in a Higgs field. As explained in chapter ten, this field exists in the space separating the surfaces of VE-spheres converging on a point of encounter. A Higgs field generates the mass of elementary particles. In the proposed scenario, therefore, mind events occurring in the vicinity of particles influence the inertial mass of those particles. As explained above, the experiment should show that the average curvature of particle trajectories in a cloud chamber is decreased.

The emotions of human participants in the experiments using a cloud chamber could be produced by shocking or pleasant images projected on a screen. To maximize the number and intensity of mind events created by these images and by other sensory input occurring simultaneously, the participants could be told that the experiment is to test the ability of attentive, multi-sensory observations. To ascertain the effect size of cognitive efforts, simple algebra problems comprehensible at a glance could be projected on the screen, and the participants could be told that they should jot down answers and observations on a questionnaire handed to them beforehand, and that the questionnaire will be graded.

The proposed experiment would show whether the strength of the hidden influence propagated by the mind events of a necessarily limited number of participants is large enough to cause detectable changes of the random events in a cloud chamber. The underlying assumption that random events can be influenced by mind events without specific intentionality is supported by the experiments designed by Roger Nelson and described at the beginning of chapter eight. Random number generators installed at locations around the world reportedly showed signification deviations from random data when spectacular events attracted worldwide attention.

The existence of a *global field of consciousness* suggested by Nelson's experiments and the assumption that consciousness may influence random events irrespective of their cognitive content or intention dovetails with the effects hypothesized by the Physics of Encounter. Additional experiments could determine whether the spectacular world events that change the output of Nelson's random number generators also influence the events in cloud chambers as described above.

Experiments relevant in this context were also devised by Haisch and Rueda. They reported a correlation between the *acceleration* of a detector system and "electromagnetic zero-point fluctuations" (16/3). An experiment using the same kind of detector system should show that analogous changes in the fluctuations will be caused by a large enough number of mind events occurring in close proximity to the detector system when it remains *stationary*.

Experiments with particle accelerators have yielded results that seem to confirm various aspects of the proposed scenario. One aspect, in particular, could easily be verified with a greater degree of certainty if the use of animals as described above proves to be efficacious. The Physics of Encounter hypothesizes that when protons and positrons collide in a particle accelerator, the number of positrons that "rebound" into the direction from which they came will exceed the number that is calculated by using the standard model if a large number of mind events occur in the vicinity of the accelerator. The hypothesis is based on the scenario that the effect of positive charges is exerted by hemispheres that diverge from the observer in the dimension of time.

An experiment with protons and positrons that yielded an unexpectedly high number of such positrons was conducted at the HERA particle accelerator in Hamburg, Germany. As reported by Frank Grotelüschen in the daily *Berliner Zeitung*, a total of eleven inexplicable positron events occurred over a period of two years instead of the expected one or two.(1) The purpose of the above experiment, of course, had nothing to do with the mind/matter model proposed here, therefore the number of persons present when the accelerator was in operation was not ascertained. The perplexing results of the experiment, as interpreted here, were caused by the mind events of those persons.

With additional experiments conducted in the presence of a large number of animals housed in the vicinity of the accelerator, it might be possible to "fine-tune" the verification of the hypothesis by varying the number of animals. As explained in chapter thirteen, the VE-surfaces created by mind events differ from the VE-surfaces that expand from the atoms of inanimate matter. One of the distinguishing characteristics is the imbalance of the spin effects created by VE-surfaces expanding from the atoms of matter that is alive. Another difference is the role of up quarks and the positive charge associated with them. As exemplified by the weight difference between protons and neutrons, a positive charge reduces the inertial mass of particles containing quarks. The entropy-inhibiting effect of left-handed spin in protein molecules and the role of electric charges in biological matter were described in chapter thirteen.

Experimental evidence that supports the proposed scenario is the transient weight increase of animals at the moment of death. This phenomenon has been investigated by Hollander (13/12) and others. Corroborating evidence is also provided by weight changes of PK agents in séances during the levitation of objects (10/14) and during the creation of rapping sounds (10/15), as described by Pilkington, and by the weight changes of RSPK agents during Poltergeist events, as described by Roll (10/16).

Another hypothesis that lends itself to verification by experiments with particle accelerators is the creation, from hadron collisions, of short-lived hybrid particles manifesting characteristics of quarks and of leptons. Experiments at the above-mentioned accelerator in Hamburg unexpectedly produced particles that seemed to possess those characteristics. They were tentatively called *leptoquarks*. More research is needed to determine whether such particles are identical with the elusive "suzy" particles. If their existence is reliably confirmed, Grotelüschen explained, this would support the hypothesis of *supersymmetry*.(2)

The existence of leptoquarks follows from the proposed scenario of particle creation and decay. As described in chapter twenty-one, first generation quarks and leptons are created within one effect plane at a point of encounter between the expanding circumferences of VE-spheres. At the site where encounters cause a hadron to disintegrate, the surface of a quark that has not been encountered continues to expand. Both quarks and electrons have a surface and a core (a point of origin). The core is the initial burst of vacuum energy released at a point of encounter. The expanding surface is energized by the quantum vacuum fluctuations of the ambient space. In the case of quarks, the surface corresponds to what Steve Geer of the Fermi National Accelerator Laboratory, according to the science journalist Thomas de Padova (11/1), called the outer shell of a quark. Since particle acceleration increases the number of encounters between VE-spheres, the size of the VE-spheres and of the particles involved in the encounters decreases. If the VE-spheres are exceedingly small, the core energy and the surface energy of leptons and quarks become indistinguishable and the particle seems to be a "leptoquark".

Experiments using the particle accelerator "Tristan" in Japan have confirmed the nature of particle surfaces as described here. Electrons, which normally repel one another, were "pushed together" in the particle accelerator. As reported by Grotelüschen in *Süddeutsche Zeitung*, the experiment showed that "an electron is wrapped in a mysterious garb of 'virtual' particles". The results, according to Grotelüschen, were published by David Koltick in *Physical Review Letters*, Vol. 78, 1997.(3) The 'virtual' particles, as interpreted here, are the effects created at points of encounter between the expanding QV-bubbles that energize the expansion of a VE-surface. The surfaces of particles, in the scenario proposed by the Physics of Encounter, are the surfaces of interacting VE-spheres.

As of this writing, the most ambitious experiment with a particle accelerator ever attempted in the history of physics is still months away. After a failed start-up in 2008, the project at the European Nuclear Research Center (CERN) near Geneva was rescheduled for 2009. As described by Aaron E. Hirsh in the *International Herald Tribune*, "the Large Hadron

Collider, Europe's $9 billion investment in particle physics, will take a handful of ions, hurl them through 17 miles of circular tunnel and smash them together so hard that they will shatter into the finest atomic shards anyone has ever observed. If all goes according to plan, the glints and flashes from those shards will at last reveal the mysterious Higgs boson, the one particle that endows all others with the property of mass."(4)

I will go out on a limb here and state that this particle does not exist in the observable part of our universe. Its location is the point of encounter where a *Higgs field* originates, and that encounter occurs in the undetectable substructure of space that theoretical physicists have equated with quantum vacuum fluctuations. The Higgs boson, as interpreted here, cannot be adequately described by the standard model of mainstream physics. In the proposed scenario of mind/matter interaction, the Higgs boson exists at the "other end" of oscillating strings converging on a common point of encounter. The metaphorical strings connect the observable universe with an unobservable substratum of physical reality. The oscillating thrusts of the "strings" are exerted toward the observer and away from the observer in the dimension of time.

The unobservable part of our universe contains the interacting QV-bubbles that are the quantum vacuum fluctuations of space. The fluctuations occur as the QV-bubbles in the surfaces of VE-spheres expand and collapse. The space between VE-surfaces converging on a point of encounter becomes a Higgs field when the encounter occurs. As described in chapter ten, the convergence involves a cascade of encounters between the QV-bubbles in the converging surfaces. The Higgs boson is created when the cascade ends. The observer participates in the creation of the cascade, i.e. in the creation of a Higgs field. As defined by the Physics of Encounter, the "location of the observer" is the location of the QV-bubble that contains the qualia of the observer's consciousness. The qualia are created within the QV-bubble that participates in the encounter.

VE-surfaces contain countless layers of QV-bubbles, i.e. superpositions. Within a specific superposition, all QV-bubbles contain identical qualia of consciousness. Since a VE-sphere instantly expands across what may be a large segment of space, the location of an observer within a VE-surface is not a location in space, but a point in time. The Higgs boson is unobservable because it is located at a *different* point in time. The QV-bubble that participates in the creation of a Higgs boson and the VE-surface that contain the qualia of the observer's consciousness are separated by the length of an oscillating string.

The Higgs boson is located at the common point of encounter of countless oscillating strings. As explained in chapter three, the thrusts of the strings converge on that point from all directions of space and create the dimension of time. Time is a *process* that is energized by the oscillation of "strings", or, to use the other metaphorical image described in this book, by the interaction of the QV-bubbles in the surfaces of VE-spheres.

The Higgs boson is the "particle" that creates all *emergent reality*. It is not only the "particle that endows all others with the property of mass". It is an infinitesimal portion of the energy contained in what the astrophysicist Günther Hasinger (4/8) called the "spark that ignited the big bang". The spark was the effect of a collision that took place in the unobservable "other half" of our universe, i.e. on the other side of the "temporal divide" described by the astrophysicist Paul Davies (19/4). The collision occurred *before* the explosion that propelled all observable matter into the universe we inhabit. "We live on a brane", as Stephen Hawking said (19/5), and we cannot see what he called the "shadow brane". The Higgs boson is located on the shadow brane.

The two branes are not worlds apart. They are intimately "entangled". They are connected by the thrusts of oscillating strings. As part of the spark that ignited the big bang, the Higgs boson is the energy that ignites what Michael Manthey (4/9) called the "bit bangs" in our universe. As interpreted here, the elementary constituents of information (bits) are created at interrelated points of encounter between the expanding surfaces of VE-spheres.

Information is an integral part of the mind that processes it. The Higgs boson, in the proposed scenario, is unobservable as a physical reality outside our own mind. It exists in the "other half" of our universe, and that reality exists within our consciousness. It is the physically undetectable universe within each conscious mind, inaccessible to acts of observation by others.

To underscore the conviction of mainstream physicists that the Higgs boson is the key to understanding the structure of our universe, the Nobel Prize physicist Leon Lederman (10/11) called it the "God Particle". Lederman's recourse to a conceptual construct from the domain of religion aptly illustrates the spiritual implications of the efforts by scientists to grapple with the ineffable reality of what John Jahn and Brenda Dunne simply called the "Source". As interpreted here, it contains not only the Higgs boson that endows matter with mass. The reality that contains the "God Particle" enables matter to become aware of its own existence as it evolves into a conscious human mind, imbued with the desire to understand the reality of which it is a part.

*Matter is filled with curiosity.*
Richard Feynman

Jahn and Dunne have "proposed that our palpable physical surround is an emergent property of a much vaster intangible reservoir of potential information, which we have labeled the Source, and that the emergence is enabled by the resonant coupling of this Source with its cosmic complement, the organizing Consciousness".(5) This corresponds to the characteristics of what I have described as the quantum vacuum fluctuations of interacting QV-bubbles containing the qualia of consciousness. The "resonant coupling" corresponds to the term "morphic resonance", coined by Sheldrake, which I have used in this book.

Theoretical physicists have established that a quantum vacuum seethes with interacting "virtual particles". In the scenario proposed here, the virtual particles in the quantum vacuum of a QV-bubble include Higgs bosons. Their intangible reality exerts equally intangible effects into the ambient space. The effects are transmitted by the expanding surfaces of VE-spheres. Higgs bosons are part of the unobservable reality that energizes our conscious mind as well as its "cosmic complement, the organizing Consciousness" described by Jahn and Dunne. As interpreted here, the latter encompasses what Jung called the "collective unconscious".

Higgs bosons are indestructible constituents of what I have called the gluonic light of consciousness in the nuclei of atoms interacting within an electromagnetic wave. Since the surfaces of all elementary particles contain the vacuum energy in QV-bubbles, Higgs bosons are part of the "mysterious garb of 'virtual' particles" that characterizes the surfaces of electrons when they are pressed together in a particle accelerator, as reported by Grotelüschen (22/3).

Even though Higgs bosons do not exist as individually detectable particles, they are part of a larger, intangible whole that contributes to the creation of an energy flux. Such an event is analogous to the unpredictable materialization of a particle, in a coupled map lattice, through the morphic resonance between intangible processes. By contributing energy to these processes, Higgs bosons contribute to the "glints and flashes" detectable in a particle accelerator, as described by Hirsh (22/4), but such flashes of energy should not be attributed to a single, detectable particle. They are the accumulated energy of countless Higgs bosons created by encounters within the expanding surfaces of VE-spheres.

The spectacular experiment at the CERN facilities in Switzerland was designed to confirm the standard model of elementary physical reality by providing evidence that the Higgs boson is *real*, not just a debatable conceptual construct. The final verdict is not yet in, and may not be for some time to come. Some of the hypotheses proposed by various dissenting voices will probably be validated or falsified by experiments at some future date, when the requisite experimental technology has been developed. Physicists had to wait many years, for example, until the state of the art made it possible to verify the EPR-paradox, which had been predicted on the basis of purely theoretical considerations.

The arguments presented in this book have shown that consciousness-related anomalous phenomena, from UFOs to psychokinesis and NDEs, can be explained by the standard model if its conceptual constructs are interpreted in terms of the proposed model of mind/matter interaction. It is a metaphorical model and is not mathematically formalized. But the following argument by Jahn and Dunne is relevant here. "No, metaphor is not a sloppy form of conceptual representation; it is a critical step in establishing the foundations of any objective science (...). In fact, the implicit reliance of objective science on metaphor as a means of sensorial association will need to be elevated to a more explicit functional role, wherein the commonalities of (...) experiences can be assembled into an interdisciplinary (...) structure of (...) cause and effect."(6)

In the final analysis, the hypothesized relationship of cause and effect must be expressed in quantifiable terms that, in principle, can be verified by experiment. In the scenario proposed here, verification is limited to quantifiable *probabilities* because the relationship between consciousness-related influences and physical phenomena is predicated on the random occurrence of morphic resonance between intangible realities.

The "sensorial association" mentioned by Jahn and Dunne plays an important role not only in the search for appropriate metaphors. Scientific endeavors activate human sensibilities and can be a source of distinctly Epicurean pleasure.

306

*The scientist does not defy the universe.  He accepts it.*
*It is his dish to savor, his realm to explore;*
*It is his adventure and never-ending delight.*
*It is complaisant and elusive but never dull.*
*It is wonderful both in the small and in the large.*
*In short, its exploration is the highest occupation for a gentleman.*

I. I. Rabi

## NOTES

(1) Grotelüschen (August 1997), p. 7
(2) Ibid, p. 7
(3) Grotelüschen (March 1997), p. 26
(4) Hirsh (2009), p. 6
(5) Jahn and Dunne (2004), p. 567
(6) Jahn and Dunne (1997), p. 218

# REFERENCES

Almeder, Robert (1996) Recent Responses to Survival Research. *Journal of Scientific Exploration*, Vol. 10, No 4.

Angier, Natalie (2008) Intuition and math: a powerful correlation. *International Herald Tribune,* September 18.

Aranov, R.A. (1971) On the foundations of the hypothesis of the discrete character of space and time. *Time in Science and Philopsophy.* Slovak Academy of Sciences, Elsevier Publishing Company, Amsterdam, London, New York.

Arp, Halton (2000) What Has Science Come to? *Journal of Scientific Exploration*, Vol. 14, No. 3.

Arp, Halton (2007) Letters to the Editor. *Journal of Scientific Exploration*, Vol. 21, No. 3.

Atmanspacher, Harald, *et al* (2006) From the Dynamics of Coupled Map Lattices to the Psychological Arrow of Time. *American Association for the Advancement of Science, Pacific Division,* 87th Annual Meeting in San Diego, California, June 18-22

Beichler, James (2007) The Paraphysical Consequences of 4-D Space. 26th Annual Meeting of the *Society for Scientific Exploration,* May 30-June 2, in East Lansing, Michigan

Bierman, Richard J. (2006) Empirical Research on the Radical Subjective Solution of the Measurement Problem. Does Time Get its Direction through Conscious Observation? 87th Annual Meeting of the *American Association for the Advancement of Science, Pacific Division*, San Diego, California, June 18-22

*Bild-Zeitung* (1990) Der Kugelblitz überholte mich auf der Treppe. Hamburg, Germany, October 17.

Boes, Roderick (2005) The Physics of Encounter. *Society for Scientific Exploration*, 24th Annual Meeting in Gainesville, Florida, May 19-21.

Bohm, David, Hiley, B.J., and Kaloyerou, P.N. (1987) *Physics Reports*, Vol. 144.

Bonvin, Fabrice (2005) *Ovnis: Les agents du changement.* JMG éditions, Agniéres, France.

Brown, Courtney (2007) Toward a Quantum Mechanical Interpretation of Collective Consciousness. *Society for Scientific Exploration*, 26th Annual Meeting in East Lansing, Michigan, May 30-June 2.

Chang, Kenneth (2009) Stellar upgrade: Milky Way bigger than we thought. *International Herald Tribune*, January 7.

Cramer, John G. (2006) Reverse Causation and the Transactional Interpretation of Quantum Mechanics. *American Association for the Advancement of Science, Pacific Division*, 87th Annual Meeting in San Diego, California, June 18-22. faculty.washington.edu/jcramer/PowerPoint/AAAS_20060621%20.ppt

Cramer, John G. (2006) *Personal Communication*. cramer@phys.washington.edu, July 12.

Davies, Paul (1978) *The Runaway Universe*, J.M. Dent & Sons, Ltd, London

*Der Spiegel* (1986, Nr. 3) Welt aus Blasen. Hamburg, Germany.

*Der Spiegel* (1994, No. 19) Wunder im Verborgenen. Hamburg, Germany.

*Der Spiegel* (1997, No. 33) Feurige Blitze. Hamburg, Germany.

*Der Spiegel* (1997, No. 51) Kreisbahn der Schöpfung. Hamburg, Germany.

*Der Spiegel* (2008, No. 17) Register. Hamburg, Germany.

Diedrich, Kurt (2006) Neues vom Brummton? *Deguforum*, Bad Kreuznach, Germany, Vol. 13, No. 51, September.

Dworschak, Manfred (2009) Motor des Lebens. *Der Spiegel* No. 24, Hamburg, Germany.

von Eickstedt, Egon (1954) *Atom und Psyche*. Ferdinand Enke Verlag, Stuttgart, Germany.

Einstein, Albert, Podolsky, B., and Rosen, N. (1935) *Physical Review*, Vol. 47.

Elitzur, Avshalom C. (2006) Retrocausal Quantum Measurement: Some New Findings and Their Interpretation. *American Association for the Advancement of Science, Pacific Division*, 87th Annual Meeting in San Diego, California June 18-22.

Folger, Tim (2002) Does the Universe Exist if We're Not Looking? *Discover*, Vol. 23, No. 6.

Folger, Tim (2003) At the Speed of Light. *Discover*, Vol. 24, No. 4.

Gentes, Lutz (1996) *Die Wirklichkeit der Götter.*
bettendorf'sche Verlagsanstalt, Munich, Germany.

Görlitz, Axel, and Pfau, Tilman (2001) Kalte Atome im Kollektiv.
*Neue Zürcher Zeitung*, Zurich, Switzerland, December 5.

Goldberg, Carey (2005) Empathy may begin at the neurons.
*International Herald Tribune*, December 15.

Greene, Brian (2000) *The Elegant Universe.* Vintage Books, New York.

Grotelüschen, Frank (1997) Des Elektrons neue Kleider.
*Süddeutsche Zeitung*, March 13, Munich, Germany.

Grotelüschen, Frank (1997) Die Revolution der Physik
wird erneut verschoben. *Berliner Zeitung*, August 6, Berlin, Germany.

Haisch, Bernard, and Rueda, Alfonso (2001) A Quantum Vacuum Base
for Gravitation. *Society for Scientific Exploration*,
20th Annual Meeting in La Jolla, California, June 7-9.

Hagelin, John (2007) The Non-Local Mind – the Effects of Group
Meditation on Crime, Terrorism and International Conflict.
*Society for Scientific Exploration*, 26th Annual Meeting
in East Lansing, Michigan, May 30-June 2.

Hasinger, Günther (2003) Urknall als Einstiegsdroge.
Interview by Olaf Stampf published in *Der Spiegel*, No. 49.

Havemann, Robert (1964) *Dialektik ohne Dogma?*
Rowohlt Taschenbuch Verlag, Reinbek bei Hamburg, Germany.

Hawking, Stephen (2001) *Das Universum in der Nußschale.*
Hoffmann und Campe Verlag, Hamburg, Germany.

Hein, Simeon (2007) Anomalous Energies and Balls of Light
in Crop Circles: Photonic Refraction, Spontaneous Crystallinity, and
Quasi-Particles. *Society for Scientific Exploration*, 26th Annual Meeting
in East Lansing, Michigan, May 30-June 2.

Hirsh, Aaron E. (2009) A new kind of Big Science.
*International Herald Tribune*, January 16.

Hollander, Lewis E., Jr. (2001) Unexplained Weight Gain Transients at the Moment of Death. *Journal of Scientific Investigation*, Vol. 15, No. 4.

Hotson, Donald L. (2002) Dirac's Equation and the Sea of Negative Energy. *Infinite Energy*, Vol. 8, Issue 43, May/June.

Ibison, Michael (2006) Are Advanced Potentials Anomalous? 87th Annual Meeting of the *American Association for the Advancement of Science, Pacific Division,* June 18-22, in San Diego, California

Ibison, Michael (2007) Letters to the Editor. *Journal of Scientific Exploration*, Vol. 21, No. 3.

*International Herald Tribune*, Reuters (2000) Geneva Atomic Lab Extends Research. September 15.

*International Herald Tribune (2008)* Accelerator Anxiety. September 13-14.

Jahn, Robert G. and Dunne, Brenda J. (1997) Science of the Subjective. *Journal of Scientific Exploration*, Vol. 11, No. 2.

Jahn, Robert G. and Dunne, Brenda J. (2004) Sensors, Filters, and the Source of Reality. *Journal of Scientific Exploration*, Vol. 18, No. 4.

Jahn, Robert G. *et al* (2007) Response of an REG-Driven Robot to Operator Intention. *Journal of Scientific Exploration*, Vol 21, No. 1.

Jung, Carl G. and Pauli, Wolfgang (1955) *The Interpretation of Nature and the Psyche.* Bollingen Series LI, Princeton University Press.

Kast, Bas (2002) Die Zeit for dem Urknall. *Der Tagesspiegel*, Berlin, Germany, April 26.

Kast, Bas, and Nestler, Ralf (2008) Was vom Urknall übrig blieb. *Der Tagesspiegel*, October 8, Berlin, Germany.

Ketterle, Wolfgang (2001) Atome im Gleichschritt. *Der Spiegel* No 42, Hamburg, Germany.

Klein, Stefan (1998) Die Welt aus dem Nichts. *Der Spiegel* No. 52, Hamburg, Germany.

Kugel, Wilfried (1996) Grundzüge einer relativistischen Theorie der Psi-Phänomene. *Zeitschrift für Parapsychologie und Grenzgebiete der Psychologie*, Jahrgang 38, Nr. 3/4, Freiburg i. Br., Germany.

Laszlo, Ervin (1996) Subtle Connections: Psi, Grof, Jung, and the Quantum Vacuum. *The International Society for the Systems Sciences and The Club of Budapest,* DynaPsych Table of Contents.

Lederman, Leon (1993) *The God Particle.* Houghton Mifflin Co., Boston, New York.

Lemonick, Michael D. (2005) Science on the Fringe. *Time* Magazine, May 30.

Leonhard, Walter (1974) Neue Theorie über Entstehung des Weltalls. *Der Tagesspiegel,* February 23, Berlin, Germany.

Lorimer, David (1990) *Whole in One.* Penguin Books, London.

von Lucadou, Walter (1997), *Psi-Phänomene.* Insel Verlag, Frankfurt am Main, Germany.

von Ludwiger, Illobrand (1992) *Der Stand der UFO-Forschung.* Verlag Zweitausendeins, Frankfurt/Main, Germany.

Maccabee, Bruce (1999) Optical Power Output of an Unidentified High Altitude Light Source. *Journal of Scientific Exploration, Vol. 13, No. 2.*

Mack, John E. (1994) *Abduction.* Charles Scribner's Sons, New York.

MacLaine, Shirley (1983) *Out on a Limb.* Bantam Books, New York.

Manthey, Michael (1997) A Combinatorial BIT Bang Leading to Quaternions. *1997 Helsinki Conference on Emergence, Complexity, Hierarchy, Organization.* arXiv:quant-ph/9809033v1, September 11, 1998 (sic).

Misner, C. W. and Wheeler, J. A. (1957) Classical Physics as Geometry. *Annals of Physics* 1957/2.

Moddel, Garret (2004) Entropy and Subtle Interactions. *Journal of Scientific Exploration,* Vol. 18, No. 2.

Moddel, Garret, and Walsh, Kevin (2007) A Very Simple Test for Psi. *Society for Scientific Exploration,* 26th Annual Meeting in East Lansing, Michigan, May 30 – June 2.

Moga, Margaret, and Bengston, William F. (2007) The Possible Importance of Frequency and Duration of Treatment in Bioenergy Healing. *Society for Scientific Exploration*, 26th Annual Meeting in East Lansing, Michigan, May 30-June 2.

Nelson, Roger D. (2002) Coherent Consciousness and Reduced Randomness: Correlations on September 11, 2001. *Journal of Scientific Exploration*, Vol. 16, No. 4.

Overbye, Dennis (2008) The end is nigh! A big stakes suit to save us all. *International Herald Tribune*, March 31.

de Padova, Thomas (1996) Wohin kein Auge reicht. *Der Tagesspiegel*, Berlin, Germany, February 8.

de Padova, Thomas (1998) Das Zeitlose in der Mikrophysik. *Der Tagesspiegel*, Berlin, Germany, December 14.

*Phänomene* (1994), Karl Müller Verlag, Erlangen, Germany. By permission of Orbis Publishing, Ltd.

Pilkington, Rosemarie (2006) *The Spirit of Dr. Bindelof.* Anomalist Books, PO Box 577, Jefferson Valley, New York 10535.

Popper, Karl R. and Eccles, John C. (1977) *The Self and Its Brain.* Springer International, New York.

*Popular Science*, Vol. 9 (1968) Grolier Incorporated, New York.

Puthoff, Harold E. (1996) CIA-initiated Remote Viewing Program at Stanford Research Institute. *Journal of Scientific Exploration*, Vol. 10, No. 1.

Puthoff, Harold E. (2001) Einstein's Legacy – Will it Endure? *Society for Scientific Exploration*, 20th Annual Meeting in La Jolla, California, June 7-9.

Puthoff, Harold E. (2002) *Personal Communication.* Puthoff@aol.com, August 23.

Puthoff, Harold E. (2003) SRI Remote Viewing: Classified Beginnings. *Society for Scientific Exploration*, 22nd Annual Meeting in Kalispell, Montana, June 12-14.

Rein, Glen (1995) The In-vitro Effect of Bioenergy on the Conformational
States of Human DNA in Aqueous Solutions.
*Journal of Acupuncture and Electrotherapeutics*, Vol. 20.

Ridley, Kim (2007) Native Intelligence. *Ode*, New York, NY, May.

Ring, Kenneth (1984) *Den Tod erfahren - das Leben gewinnen.*
Scherz Verlag, Bern, Switzerland; Munich, Germany; Vienna, Austria.

Rodeghier, Mark (2007) UFO Research: Where are We Now?
*Society for Scientific Exploration,* 26th Annual Meeting
in East Lansing, Michigan, May 30-June 2.

Roll, William G. (2003) Poltergeists, Electromagnetism and Consciousness.
*Journal of Scientific Exploration*, Vol. 17, No. 1.

Rubik, Beverly (2002) The Biofield Hypothesis: Its Biological Basis
and Role in Medicine.
*Journal of Alternative and Complementary Medicine*, Vol. 8, No. 6.

Scharf, Rainer (2001) Teilchen in eisiger Falle.
*Frankfurter Allgemeine Zeitung,* Frankfurt, Germany, October 10.

Schnabel, Ulrich, and Rauner, Max (2001) Physik am Tiefpunkt.
*Die Zeit*, Hamburg, Germany, October 11.

Schuster, Gerd (1989) Mais, wie er singt und klagt.
*Stern*, March 16, No. 12, Hamburg, Germany.

Schwartz, Gary E. R., and Russek, Linda G. S. (1999) Registration of Actual
and Intended Eye Gaze: Correlation with Spiritual Beliefs
and Experiences. *Journal of Scientific Exploration*, Vol. 13, No. 2.

Sheldrake, Rupert (2000) *Dogs That Know When Their Owners
Are Coming Home.* Arrow Books, London.

Simon, Claus Peter (2004) Editor's introductory remarks (no title).
*Geowissen*, No. 33, March 15, Hamburg, Germany.

Smolin, Lee (2006) *The Trouble with Physics.* Houghton Mifflin, Boston.

Snow, Chet B. (1991) *Zukunftsvisionen der Menschheit.*
Ariston Verlag, Geneva and Munich, Germany.
(Mass Dreams of the Future, 1989, McGraw Hill, New York)

Stahlhut, Christian (1993) Die Nah-Todeserfahrungen.
*Sender Freies Berlin*, manuscript of radio broadcast, September 19.

Stampf, Olaf (1998) Der erschöpfte Schöpfer.
*Der Spiegel*, No. 52, Hamburg, Germany.

Stirn, Alexander (2003) Was den Kugelblitz zusammenhält.
*Süddeutsche Zeitung*, Munich, Germany, October 16.

Swartz, Tim (2007) As Time Goes by: UFO's and Time Distortion.
*Flying Saucer Review*, Spring.

Tegmark, Max (1999) The Importance of Quantum Decoherence
in Brain Processes. quant-ph/9907009v2, November 10.

*Time* Magazine (1984) An E.S.P. Gap. January 23.

Tucker, Jim B. (2008) Ian Stevenson and Cases of the Reincarnation Type.
*Journal of Scientific Exploration*, Vol. 22, No. 1.

Turner, David J. (2003) The Missing Science of Ball Lightning.
*Journal of Scientific Exploration*, Vol. 17, No. 3.

Vaas, Rüdiger (1998) Stephen Hawkings Weltmodell.
*Bild der Wissenschaft*, No. 5, Stuttgart, Germany.

Vallee, Jacques (1969) *Passport to Magonia*.
Contemporary Books, Chicago.

Vallee, Jacques (1998) Estimates of Optical Power Output in Six Cases of
Unexplained Aerial Objects with Defined Luminosity Characteristics.
*Journal of Scientific Exploration*, Vol. 12, No. 3.

VanWijk, R. (2001) Bio-Photons and Bio-Communication.
*Journal of Scientific Exploration*, Vol. 15, No. 2.

Veneziano, Gabriele (2004) Die Zeit vor dem Urknall.
*Spektrum der Wissenschaft*, August. Heidelberg, Germany.

Vjatkin, Lev (1993) My Encounter with a UFO. *Aura-Z*, No. 1, March.
From 1994 Newsletter of MUFON-CES.

Wendt, Hans (2007) *Personal communication*.
hans.wendt@comcast.net, November 3.

Wolf, Fred Alan (1994) *The Dreaming Universe*. Touchstone, New York.

Zaleski, Carol (1987) *Otherworld Journeys.*
Oxford University Press, New York.

Zimmer, Ernst (1964) *Umsturz im Weltbild der Physik.*
Deutscher Taschenbuch Verlag, dtv, Munich, Germany.

Zips, Martin (2005) Ein Schreck im Kornfeld.
*Süddeutsche Zeitung*, Munich, Germany, November 16.

Zizek, Slavoj (2006) *The Parallax View.*
The MIT Press, Cambridge, Massachusetts.

Zukav, Gary (1979) *The Dancing Wu Li Masters.*
William Morrow and Company, New York.